Capitol Hill
I0609182

praise for

Lancelot Schaubert

&

Bell Hammers

"BELL HAMMERS is written in a style not unworthy of John Kennedy Toole and William Faulkner – the vivid characterization of Southern ethnography commingled with stark, episodic spectacle breathes with the spirit of quintessential Americana. It is a text I would happily assign in an American Novel class and would expect it to yield satisfying discourse alongside works in the canon, whether beside the sardonic prose of Mark Twain or the energetically painful narratives of Toni Morrison."

— Dr. Anthony Cirilla

"BELL HAMMERS is the kind of story that makes you a better person and stays with you long after you put it down."

— F.C. Shultz,
author of The Rose Weapon

"Loved BELL HAMMERS because Lancelot wrote about people who don't get written about enough and he did it with humor, compassion, and heart."

— Brian Slatterly,
author of Lost Everything and editor
of The New Haven Review

"Schaubert's words have an immediacy, a potency, an intimacy that grab the reader by the collar and say, 'Listen, this is important!' Probing the bones and gristle of humanity, Lancelot's subjects challenge, but also offer insights into redemption if only we will stop and pay attention."

— Erika Robuck, national bestselling
author of Hemingway's Girl

"Myth, regret, the lore of our heritage and the subtle displays of our castes — no one so accurately and imaginatively captures the joys and sorrows of life in the Midwest as Schaubert does here. BELL HAMMERS is a Tree Grows in Brooklyn as told by Gabriel Garcia Marquez if Marquez lived in rural Illinois and only told stories to his grandkids. Seriously a delight to read."

— Colby Williams,
author of the Axiom Gold Medal winning book
Small Town, Big Money

"I'm such a fan of Lancelot Schaubert's work. His unique view and his life-wisdom enriches all he does. We're lucky to count him among our contributors."

— Therese Walsh,
author of The Moon Sisters and Editorial
Director of Writer Unboxed

"Lancelot Schaubert writes with conviction but without the cliché and bluster of the propaganda that is so common in this age of blogs and tweets. Here is a real practitioner of the craft who has the patience to pay attention. May his tribe increase!"

— Jonathan Wilson Hartgrove,
author of Common Prayer and
The Awakening of Hope

"Lancelot's attentive, thoughtful, a bit quirky, and innovative. He continues to impress me. He 'sees,' and BELL HAMMERS is full of details that enable his audience to see. Bravo, Lance."

— Jackina Stark,
author of Things Worth Remembering
and Tender Grace

"Schaubert's narratives are emotionally stirring with both a vulnerable sensibility and rawness to them. BELL HAMMERS will take you on a journey full of open wounds, intimate successes and personal delights. Lancelot's words have a calmness, a natural ease but the meaning is always commanding and dynamic."

— Natalie Gee,
Brooklyn Film Festival

"Much to admire in this story."

— *The New Yorker*
fiction editor on an excerpt

"Wonderful... honestly a good story."

— *Glimmer Train*
Finalist in their last ever fiction open

"Lance Schaubert is a dear friend, which makes me hesitant to write a blurb.... "*of course you'll make him sound good, you know him,*" you might say. But the truth is, I'd recommend this novel even if I'd never heard of the man, with everything I possess. I've told Lance when his work isn't great; now believe me when I tell you that this IS.

"This novel is true Americana; honest and even brutal in its depiction of the evils that have always haunted us, but pure and funny in its representation of the Midwest blue collar hero. It's incredibly full of heart, without ever giving way to outright pessimism or sappiness. It plays the true middle ground, bringing a quiet hero to life and even inviting God himself into the story as a character you've never seen... while making you feel that, if he's real, maybe this is who He really is.

“Lance explores the quest for justice, faith, and goodness in a way that makes you feel like you’re listening to your own grandpa—the crazy one, the one whose stories your mom was afraid to let you hear.”

— Mark Neuenschwander
Award-Winning Photographer

“One of a kind book that tells serious issues in a funny way. While reading the first chapter, I knew how special the main character Remus is. Humorous and heartfelt throughout. The relationship between Remus and relatives is very relatable. Surprisingly I found the language mimicking some classic authors. Well thought and more pressing than I thought it would be!”

— Monika’s Book Blog

“Excellent piece of writing… reminds me of Mark Twain’s works. It is sardonic at times, taking a sarcastic tone and mocking the reader while delivering an important piece of the story at the same time.”

— Scribbes

"Often humorous, there is a folkloric undercurrent to the story, as Remmy's outlook is so often painted in the shadow of his favorite fairytale, Robin Hood. Full of both humor and tragedy, I can both laugh and cry at the crazy life of Remmy Broganer. There is a palpable anxiety in the novel surrounding the polluting nature of big oil and coal, and the willingness of these executives to destroy and pollute for profit. This fight still goes on to this day. This is a fun story to read in spite of the injustices and the tragedies that seem to run in the family."

— Hana Correa, Goodreads

"In a style all of his own, Schaubert brings us the poignant history of a town, a family, that is crystal clear in minute vignettes of time and place through the eyes of youngster Wilson Remus Broganer. It is a wild ride between angst and laughter, and these protagonists are quickly included among your friends and family. You will want to read this book. This time, this place is picture-perfect and heartfelt. Schaubert is an author to follow."

—Bonnye Reed, Goodreads

"A comfortable, fun and humorous story, reminiscent of Faulkner and Twain. With loving and realistic characters and excellent writing a story that needed telling done well."

— Adventures of an Avid Reader

"A good story. Its plot is very distinctive, its themes allude to significant issues, and its narration is simple and heart-warming. This book is something completely different from the stories that I usually read, so I am grateful that I picked it up."

— Khansa Jan Dijoo

"Bell Hammers is and enjoyable and thoughtful read, one that captures life in southern Illinois Coal Country during the 20th century. The book is both funny and poignant and as I read through, I really felt myself bonding with these characters. Early in the book, Remmy is told of his grandfather's participation in the Herrin Massacre. I had never heard of this, but the telling was so real, I had to stop reading to find out if it was a real event. It was a real event, it was horrific, and I cannot believe I never heard of it in school."

— Drew K., Goodreads

"Bell Hammers follows an ornery child on his path to become an ornery man. Even from the first page the reader understands that this is no normal lead character, but one with life and stories… and pranks. The book isn't all hijinks, however. Bell Hammers also focuses on serious issues such as the ethics of big corporations. Of love and family. Of race, and the prison system. And through it all Remmy has his faith and conversations with the Lord. Overall it was a charming read that you will be thinking about long after reading the last page."

— Julie, Goodreads

"I just finished reading this. It is 12:45 AM. I couldn't stop until it was finished. It starts out kind of light and whimsical, occasionally funny. It reminds you of the stories family members would tell. Meandering. A bit slow, plodding, but nice. Nostalgic, even. And then it got really, really dark and grim, and, honestly, kind of depressing (in the way that only maybe-semi-biographical stories can be), and then it ended on a strange note. It was a good book, a very well written book, and certainly one that will stick in my memory, but wow."

— Genevieve Paquette

"Bell Hammers is a wonderfully written book that follows Remmy through life in southern Illinois. The writing is very good and reads a little like Mark Twain, especially the earlier sections. In some places, Bell Hammers reads like a series of anecdotes told at a family gathering—it was excellent."

— Ryan Mac, Kickball Champion
and Goodreads Reviewer

"I loved the language and the ambience, it was especially heartwarming to read the acknowledgements. I look forward to another Remmy story!"

— Natalie Cottingham

"A tale that takes you on a joyful ride around Egypt, Illinois. Seeing the world through Remmy's eyes is enjoying and fun. You instantly get a sense of what Remmy is about from the first chapter, and aren't let down as the years pass by. A great story that should be dipped by everyone."

— Calli, Goodreads

"This book had me hooked with its writing and character development. It made me think, smile, pause and laugh. An accomplishment only made possible by weaving the intricate tasks of good writing, timing and pacing."

— Sarah Dickinson
Author of *Silver Spoons*

"Schaubert does a great job of expressing the dialect of Southern Illinois and the chasm that exists between the laboring class and profit-focused companies."

— Dianne, Goodreads

"I read somewhere, Amazon I think, that Lance's work smacks a bit of John Kennedy O'Toole (the beloved "Confederacy of Dunces"), and a bit of John Steinbeck. Odd, because I remember thinking that exact thing before reading that someone else had invoked those names. Lance is too young to write so elegantly, so poignantly, I thought. But he is the real deal. I am a sucker for the people and places that are a part of the Bell Hammers world. I hadn't heard anybody mention Garrison Keillor, but that is apt too. Join me in watching the growing career of this extraordinarily talented writer. It should be very exciting."

— Meg Langford,
Goodreads

"I loved his witty sense of humor and his relationships with his wife and friends. I'm not religious, but I thoroughly enjoyed his conversations with God. Some of them had me cracking up."

— Maureen Mayer
author of *Relinquishing Liberty*

"Remmy is a wonderful character, set on creating a happy life for himself and the other less fortunate folk. The story is set in a region in Illinois known as Little Egypt, and describes a land of hard working farmers and oil company entrepreneurs. The style of writing is reminiscent of Mark Twain, in that the author liberally uses colloquial expression and clipped sentences. "Bell Hammers" is engaging, entertaining, and darn good distraction from all of the horrific COVID-19 news and statistics."

— Sarah Jackson
author of *Pete and the Persian Bottle*

"Bell Hammers by Lancelot Schaubert was the book I needed recently. I'd been struggling with anything I had picked up to read… until Bell Hammers."

— High School Teacher + Librarian

"In the tradition of "Predator", Plato's "symposium" and "the Hardy Boys: Secret of the Old Mill", Lance Schaubert has written a gold dream of existential steampunk romance. Again and again, I found myself delighted with the unforgettable prose, especially when it comes to the exploring the philosophy of decapitation. If you enjoy Louis L'Amour and Tolstoy, you'll find this epic western saga a delight for the brain, the heart, and of course the tingly bits."

— Mark Neuenschwander
Award-Winning Photographer

"Mark, you already did a blurb. You can't do another blurb. *Especially* that blurb."

— Lancelot Schaubert
Author of Bell Hammers, this novel

"Hi honey, so proud of you!!! 🩶🩶🩶 can I do a blurb???"

— Lance's mom
A retired nurse

"No, mom, this is… see what you started Mark?"

— Lancelot Schaubert
Author of Bell Hammers, this novel

"Mark Neuenschwander's work is a tour de force: he is the voice of his generation."

— Colby Williams,
author of the Axiom Gold Medal winning book
Small Town, Big Money

"Colby?! Mark takes *photographs*. How can he be the voice? And why are you blurbing Mark in—"

— Lancelot Schaubert
Author of Bell Hammers, this novel

"I've pranked 57 people since being inspired by the characters within and am now banned from many fine establishments including this novel."

— Mark Neuenschwander
Award-Winning Photographer

"I'm shutting this down. Right now. We have a novel to start and there's far more stake here than my ego or your… your… blurb trolling of the aforementioned."

— Lancelot Schaubert
Author of Bell Hammers, this novel

Schaubert, Lancelot
Bell Hammers / Lancelot Schaubert

ISBN-13: 978-1-949547-02-3

1. FICTION / Humorous / General
   FIC016000
2. FICTION / Family Life / Marriage & Divorce
   FIC045010
3. SCIENCE / Global Warming & Climate Change
   SCI092000

I. Salem (Illinois)
II. Little Egypt (Illinois)

Printed in the United States of America

DEDICATION

For Kiddo.

I nicknamed you, in part, because Jerry nicknamed me. And your patience with me and my craft is as great as was his family with his.

For Grandpa Jerry Schaubert.

Who used plywood as parachutes, whose Grandad wrote them big companies, and who built half the houses in Southern Illinois.

For Pawpaw Deno Bubba.

Who crashed the plane, nicknamed Mimi "Hippo Shit" after the world's largest hippo crapped on her, and trained most of the real estate brokers in Bellhammer.

For Grandpa Balu.

For giving me the idea to build Kiddo a hope chest, raising Tara gently when she got a sister, and tending to the birds of the woods.

For Opa Zeiter.

For singing to the Good Lord on your walks. Augustine said that he who sings prays twice.

For my father who's a better man than Bellhammer knows.
For my mother who's gentler than Bellhammer sees.
For the fathers that made me fight for women.
For the mothers that made me a man.
For the friends that raised me.
The siblings who refine me.
For tales dying in the nursing homes of our country.
For the children who will be born in the midst of a world in turmoil.
For the grandchildren growing up in a world whose last glaciers now melt.
For Paul M. Angle, whose book were the last two words Pawpaw Deno uttered to me and whose *Bloody Williamson* connected many of the dots in my homeland's and family's history — I have block quoted your work in the proper place and cited you in hopes that this novel, if nothing else, will serve to direct people back to your work.
For Denison Witmer — your songs and Fitzsimmons's are pretty much the only non-instrumental songs I can write to these days. I wrote most of this novel to your discography.

For Jackina Stark, Lisa Stephenson, Jeremy Redman, John McGee, Mr. Baker, Mrs. Marsh, Mrs. Lambert, and all of the other teachers who first told me to *write* and *speak* and *dance* and *sing*.

And lastly for Janie Jackson, the wonderful old Pentecostal nurse from St. John's 6 East. The one before the Joplin tornado. Who in 2009 at midnight stood next to me as a patient came in with lopped-off fingers. You said, "Honey, you damn well better not faint on me. If you gotta faint, get out." I stayed. And we mopped up a ton of blood. And then you said, "You'd better dedicate your first novel to me after all this shit, so help me God." I've never been literally up to my elbows in anything but blood, soil, and dishwater. The blood was with you that night, so I made a promise.

Blood mopped. Promise kept.

BELL HAMMERS

THE TRUE FOLK TALE OF LITTLE EGYPT

LANCELOT SCHAUBERT

1941 - LITTLE EGYPT
SOUTHERN ILLINOIS

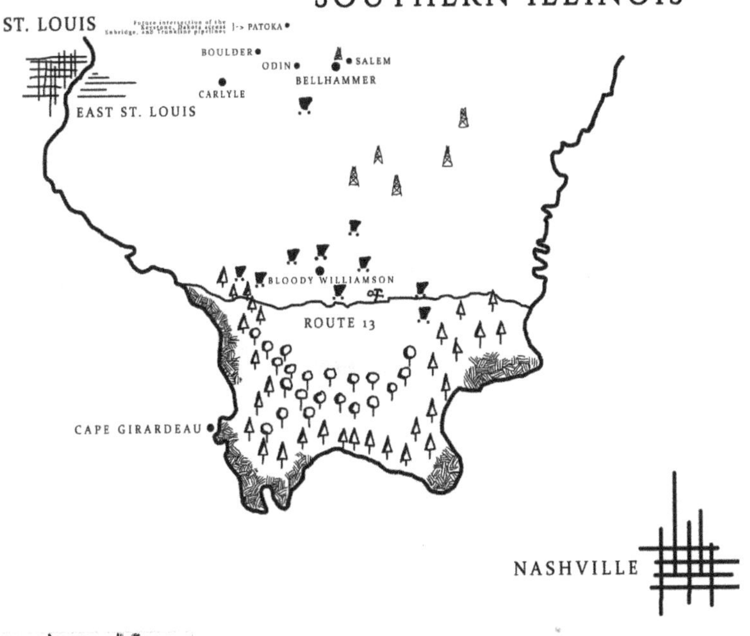

LEGEND

COAL

FOREST

ORCHARD

OIL WELL

CROP LAND

SCALE OF MILES

0 5 10 20 30

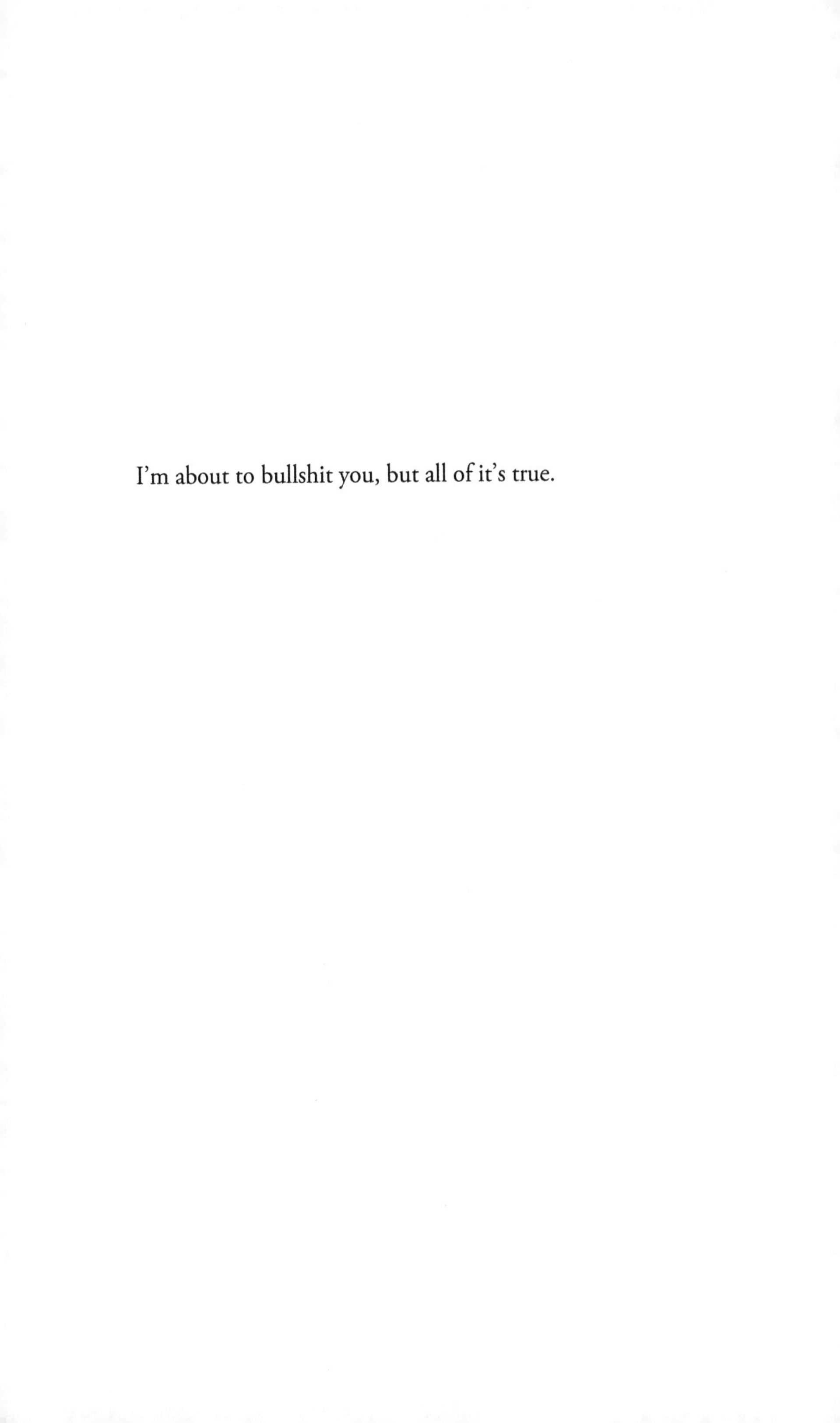

I'm about to bullshit you, but all of it's true.

A. Overture and Leitmotif

"The city is recruited from the country."

— Emerson

"The manual arts have always preceded the fine arts: some-one had to build the Sistine Chapel before Michelangelo could paint it."

— Unknown

WILSON REMUS

1941

BUCKASS NAKED IN hot, hand-boiled bathtub suds, playing with his tin New York dairy truck and some Spur Cola bottles, he heard old Rooney's brakes set to squelching.

"Aww shit." He was six years old. "Aw shitty shit shit."

They didn't have no school buses back then, you see, just one room schoolhouses dotting the countryside like peppercorns tossed sparingly over a pot of boiled taters. And if you weren't gonna walk five miles to school one way, you'd better get your ass in line for old Rooney's flatbed truck when it pulled up to your street corner when them brakes squelched out loud.

Remmy jumped up quick as a cat scared by a cucumber and ran out without drying himself. "Rooney! Rooney!" Momma Midge cried after but it was of no use.

It started to go and all of his classmates and Elizabeth too stared at him with suds all down his naked body as he sprinted across that hot dirt road and it picked up on his feet till the soles went black and he caught the truck just barely and plopped buckass naked on the back with the rest of them.

The other kids stared. One snorted.

Rooney slammed on the brakes with a fresh squelch and craned his head out the window. "The hell, Remmy?"

"The hell, Old Man Rooney?"

"Don't you *the hell* me, boy, you're buckass nekked!"

The kids giggled then. Specially Elizabeth.

Remmy blushed a bit. He was naked, but not quite old enough to be ashamed. Not quite. "So?"

"So you can't go to Sunday school nekked, Remmy!"

"You can't go to Sunday school without me, Old Rooney!"

"Well… well you're nekked though."

"Well so what? Skin and mind ain't the same."

"Don't get smart with me now. Don't you start."

"Honest, Old Man Rooney, I'd rather go to school naked than to stay home covered but dumb."

Rooney shook his head. "Go put on your britches. I'll wait." Remmy scooted off the back of that pickup and got about five feet before he heard the kids pointing and laughing. He looked down — some of the limestone dust in the back of that flatbed had stuck to his butt, and now he had a white ass to offset them black soles. Full white moon and hooves of black. Like a whitetail buck.

But they got him to class, they did. Him and the others. He sat down and tried his best to wink at Beth. He winked and he winked and fidgeted in his chair, the limestone working his buttcheeks like sandpaper.

Beth never did wink back no matter how much work Remmy'd put into winking her way. He'd give anything just to be able to fall asleep in the safety of her older, softer arms and wish the world and its scaffolding and fist fights away. Oh and its hate too, yup. But she didn't seem fond of that idea, the winking and the kissing and the holding, or even the noticing him, really, busy as she was with her maths.

Maybe she'd seen enough of him for the day, all things in mind.

Remmy'd been in the second grade at the time and learning from Miss Witt in the one-room school. Miss Witt said, "Well it looks like we got six students and four oil people today."

The children of parents not employed at Texarco laughed and pointed at the rest. The children of oil parents blushed. That included Beth.

"Missing one oil person," Miss Witt said. "Where's Jim Johnstone?"

"Probably painting himself black with tar," Remmy said.

"You quit," Beth said to Remmy.

Beth being one of them oil people put him in one of them tight spot dilemma problems, it did. Remmy went to school there along with a few other kids, learning his grammars, how to make his thoughts into clean words, but mostly just winking at Beth Donder and hoping she'd wink back.

Fat.

Chance.

She was five years older than him, which made her twelve or something. That combined with his oil people comments made it damned near impossible he'd get a wink out of her. He remembered the news came in on a Sunday morning in the middle of the Sunday school and the winking and her age.

Jim Johnstone came running in hot and sweating like a creek-dipped mink in his winter wear, that look on his face like he had bad news nobody else knew about and he'd only tell you once you begged him good and long to reveal his secrets. Except it must have been extra bad cause he said, "Miss Witt! Miss Witt! Turn on the radio!"

She turned it on.

"—C. Hello NBC. This is KTU in Honolulu, Hawaii. I am speaking from the roof of the Advertiser Publishing Company Building. We have witnessed this morning the distant view a brief full battle of Pearl Harbor and the severe bombing of Pearl Harbor by enemy planes, undoubtedly Japanese. The city of Honolulu has also been attacked and considerable damage done. This battle has been going on for nearly three hours. One of the bombs dropped within fifty feet of our KTU tower. It is no joke. It is a real war. The public of Honolulu has been advised to keep in their homes and away from the Army and Navy. There has been serious fighting going on in the air and the sea. The heavy shooting seems to be—" Static cut off the broadcast. Then the voice went silent.

The kids did too.

Remmy didn't like how quiet it was so he got up and went into the corner of the schoolhouse and dropped his britches — which showed his limestone-white ass — and started peeing in the mop bucket.

Miss Witt shouted, "Good Lord, Remmy, what on earth! Why are you doing that?"

"Cause I got good aim," he said. "Why else?"

The kids laughed.

Remmy turned his aim a bit while they was laughing and sprayed a little on Jim Johnstone's notebook just cause that boy liked being the bearer of bad news. Miss Witt sent him home early and, though happy that he made the kids laugh instead of thinking about the new war, in later years Remmy would say to me, "I couldn't believe I did that. I guess I always enjoyed the power of a good prank."

They had rationing after that. You couldn't buy sugar or coffee or gasoline or anything without a stamp, which you got from the ration board. It mattered how far you had to drive to work which messed up his Daddy John's milk jug gathering, since Daddy John had finally saved up enough to ditch the wagon and get a la bumba of a car.

Forced Daddy John to take more time building homes and sheds and things for men in the oil fields. Daddy John wasn't that close in to begin with, but Remmy hated the government for taking away his dad even further and hated Texarco for keeping him. It took away too his chance of one day having Beth to rock him to sleep safe away from shouting and wars like a good mother, curbing travel like that. See, you had to ride with somebody else wherever you went so you didn't drive so many cars. If you wore out your tires, you had to get a permit for another one — one at a time instead of a set. Couldn't get meat, so Remmy'd shoot squirrels and rabbits with his slingshot and cook them, and that's no lie.

Remmy stole stories from the one room school house — for one, cause they were expensive, books, and for another, cause boys made fun of other boys for reading and so he needed to read in private, and for a third, cause if he didn't like the book — say it tried to sound smarter than it really was deep down — and if rations got real bad, he could always use the front pages to wipe his ass.

They'd had themselves a farm — a peaceful place out away from the oil fields and out away from the milk driving, where at least one Saturday a month Remmy'd been able to play out in the yard with Daddy John. He missed the smell of that farm — the sweet corn and shitty smell of good fertile soil. But because of the travel curbing, they moved in from the farm. Moved in to the big city: Odin, Illinois. Traffic was awful when you had a twenty-four street town. They sold most of it, his parents and the farm, but they brought a couple pigs along. Them pigs was an anchor for a while, keeping Remmy joined to that heavenly garden on earth. Other people had pig pens in the back. John David — Remmy's Daddy — raised them so they could have some pork.

When the pig got turned into pork, the anchor was cut loose and he was free floating in Odin. Midge — Remmy's Momma — kept chickens so they could have those, but they weren't half the people pigs were. The chicken coops went in the side yard, and those chickens never really settled down either after the move. Remmy got it: foxes everywhere.

Shoes was hard to get all of a sudden. Hell, when he was on the farm he'd loved going barefoot, and as soon as he needed shoes to walk around town on account of moving into town on account of the war, he couldn't get good shoes also on account of the war, which wasn't fair no matter how he looked at it. Had to sole them and put heels on them over and over again, wishing he had Moses' shoes that never wore out. Couldn't buy hardly anything. So everybody dug in and did what they could do.

They had paper drives. Remmy took his paper around to people's houses and tied it in bundles and stuck it up on the wagon and sold it, hoping the money would help Daddy John not work so hard and then maybe have some time to the family. Never really worked, though. What'd they sell the paper for? Well for cardboard, for shipping crates for the war. Some of them crates had munitions, stuff for the war. Oh, yeah, they had a pants factory. Pants for the army. Cause you can't go to war with your horse running loose out of its barn, the other seven-year-olds boys all said. Specially the streakers.

Remmy had to admit that he knew something about that.

Yeah it was the big plant that'd done the bottled cola there, Spur Cola from Bellhammer, Illinois? Remmy watched that plant close one day in the war for the pants and watched them take all of those bottles — just a bunch of them — and he followed them out and saw people dump them into a specific mine shaft. Yeah, that cola plant'd shut down and turned into a place for making pants that kept the horses of the respective army men in their respective barns. That and saltpeter.

Well when they abandoned that coal mine around the same time, everybody dumped their trash down in there, down in the mine. So it seemed right when the time came to do so to lower all those full and sealed Spur Cola bottles down that shaft. Remmy watched them do it just to make room for the pants, and he was just a little boy, so he wasn't strong enough to go down in there and get them bottles, but he reminded himself of the place: the old railroad, the groundwork of the truck stop, the shoe factory, and the bottle factory near the mine. He did. Because he asked The Good Lord, "Good Lord, will you help me remember this place?"

And The Good Lord said back, "Remmy, I will. Remember me, Remmy."

And Remmy said, "Good Lord, I will."

So Remmy memorized it and The Good Lord both. Some days he'd come back and mark the spot with his toe or a flag made of a stick and a rag or write his name in the dirt there with his piss just to make sure he still knew all them bottles were hid down in there. And one day he'd come back and dig up all those bottles, cause there wasn't another Spur Cola in the world but in Bellhammer, Illinois, and therefore one day those Spur Cola bottles would be prime rare antiques, and so he'd dig up all of them and sell them one at a time on the big city auction block. A regular old Sotheby's, yes sir.

And then he'd have enough money to buy his Daddy John a vacation for just the two of them in some castle somewhere in Ireland or Germany or Camelot — somewhere where they have those old castles and throw jokes like jesters at all the dumb tyrants around the world. He wanted to build the biggest castle out of the world's greatest joke. Best part about throwing jokes and pranking tyrants is that there ain't no consequences for a good joke, and yet they change people's minds. Kind of like the joke he'd told about the castle he'd built the year before out of the Lincoln Logs in the back of the horse wagon, back when he'd gotten lost and Daddy John had shouted. That was before they'd moved in from the safety of the farm — their Little Egypt castle. Before everything went to hell and they'd treated each other like Bloody Williamson.

WILSON REMUS

1942

EVEN THE SHOESTRINGS was on a shoestring budget. For instance, you couldn't make a telephone company with two cans and a shoestring very easily, what with how much the shoestrings went for — certainly for more than the cans. So Remmy and a couple of his eight-year-old friends — Pete Taylor among them — started a real phone company.

See you can't have a Camelot and the world's biggest prank if you don't have homing pigeons so the knights can talk to one another. That's why he needed a phone company. Like a walkie talkie for army men. The lines in the county got left behind by the Army Corp of Engineers when Texarco moved on to a different field. The poles, that is, since the actual wires got sent off to fight the Germans, which was hard for most of the people in Southern Illinois to admit, particularly around Germantown, but many of them also didn't like Hitler, being Jewish of blood, or

German Catholic, so they still said they'd sent the copper to fight the Germans. And the Irish loved to say it. Well Remmy was both Irish and German. And young, so he focused instead on the telephone poles while the copper wires and the rubber tires and other things found themselves suddenly embodied and fighting The Germans.

At first they didn't know what to run betwixt the poles, seeing as how they didn't have no good wire nowhere. Then they came across an old field where the fences didn't matter, at least not to the boys, so they started pulling up barbed wire and used the barbed wire to connect the phones. They just kept taking down pieces of the fence and tying them together. They hadn't quite figured out how a bunch of eight-year-olds were going to get all that tied barbed wire up onto the poles, but they didn't mind because it gave them something to do in the summer when school wasn't as busy and their dads were off working the oil fields or working for the people who worked the oil fields or working for those people's people and so on. They'd take it one step at a time.

Around that time, this asshole named Jammie Lucas Jung-Jacques came and took over for the boys, and it was a hostile takeover if ever there was a hostile takeover in the history of telecom companies. Half the boys called him Triple J and the other half called him Jammed Jung, which meant pretty much exactly what it sounded like it meant.

Jammed Jung took up that wire and started stringing it for the boys and jolted it and told them to scat and then hooked it up to the phones in the homes of folks. And he expanded their fence destruction operation and took down pieces of fence here and there around Little Egypt and started stringing it up every which way. It was staticky — you couldn't hardly hear. But that old boy Jammed Jung got a decent chunk of change from the ideas and the work of eight year olds. That's the way it was here on the reservation.

Of course the boys had the last say. You don't steal all the homing pigeons in the columbaria of a good night of Camelot and think you're going to get away with it. So Remmy and a couple of others took to a couple of those poles with a camping hatchet. It was days and days of work for that gang of eight year olds, but they cut at it and cut at it and the thing finally fell with the great crash of an electrified redwood. All sparks and splinters. It cut out most of the service from the Triple-J Barbed Telephone Company and started a small brush fire, so that was the end of that enterprise.

Remmy saved a massive spool of that barbed wire, though. Saved it up nice. Because you never knew.

Remmy wasn't done trying to make his money to start his paradise and the world's largest prank, but for that year he focused on saving money.

You had three gasolines for your car from the Texarco pumps: "A," just to blow the dust off your windshield and down the street driving, got a "B" to go to work, and "C" for running it up to speed on the open road. You saved money by using as much A and B as your car would take, but it smelled awful. And then your book with ration stamps gave you a couple of pairs of pants a year. You'd go to the store: stamps for sugar and fruit and all kinds of stuff you couldn't get without it.

"You need me to use my power saw? Gotta charge $2/day extra." That's the way they'd do, charge extra for different and better tools on account of the war. Everything was *a la carte*. The man said, "We'll pay it." So they took the tools to the job in a little red wagon. They wasn't driving anywhere, because of the war. No gas, you see.

At Remmy's home they had those chickens and pigs and an outhouse in the alley. They didn't have water in the house. They did have electricity, but no water. Had a well and a hand pump. Had an ice box. Tired of the pan in the bottom, they ran a hole and let the water just fall out the back of the ice box. That took care of the output, the water. For input, the ice man had to get twenty-five or fifty pounds of ice in order to dish it out to everyone so that they could keep their food cold.

And that was the source of Remmy's next great money-making plan.

WILSON REMUS

1943

KIDS WOULD CHASE the ice man, jump on his stocks, and get a piece of ice chipped off. Gwen, Remmy's sister, and he would pick strawberries at the ice man's, a nickel a quart. Took six hours to pick and pick. A nickel a basket and if they stayed and if they picked all of them, they got two quarts to take home. They made a little money.

And Remmy'd hoped to earn some more money for that special place he'd build or that special time he'd set apart for him and Daddy John and fix what had happened when he'd left the horse wagon and their little Bloody Williamson happened. With a little left over maybe for candy – specially that black licorice, oh man'd he love that black licorice as much as he loved the good Lord.

He'd gone up to Mister Tolliver, who owned the general store, and had asked him if there's anything he could do to earn some money. Well Mister Tolliver'd suggested Remmy go grab a five-pound block of ice from the ice truck, and he'd give him fifty cents.

Well Remmy knew he'd lifted five pound sacks before and set them right back down. He'd even dragged bales of hay across the barn floor. So they wasn't too bad. But when he set to lugging this giant block of ice with Mr. Tolliver's grabber claw, which he'd let him use, he realized this block was something beyond his realm of humanity and moved on into the realm of the angels. The gods. Or at least the adults.

He asked The Good Lord, "Give me the strength of Samson."

And The Good Lord said, "No, Remmy, but I'll give you the long suffering of Job."

"That's a fine joke on me, Good Lord."

" But it's just the aid you need."

Took the eight-year-old an hour of hefting that thing however many miles he'd walked before he knew he'd have to set it down. It's hot and the ice's sweating more than he was, golly was it ever hot. So he set it down in the dust and thought about streams of cold water and wells that ran on forever.

By the time he'd picked it back up again, he realized the bottom would be filthy as his own was when he sat buckass naked in Rooney's truck.

Took only forty-five minutes for him to set it down again. That time, closer to town, he figured he'd better get himself a lick or three while the getting's still good, so he did. He licked that thing and, unlike some kids, his tongue didn't get stuck, luckily. He lapped some of the tiny puddles that'd pooled in the top of the block. Refreshed, he pushed himself on the next march, got himself all the way onto the edge of town, where they did the car work at the tire shop fore he had to set the blasted thing down again.

"Wilson Remus Broganer," said Henry Ferguson, from the tire shop. "You carrying that thing for Mister Tolliver?"

"I am, sir," Remmy said.

"Well I'd better tell him you're setting it down in the dirt, huh?"

"Tell'm what you want, but tell'm I didn't wait around to listen to your threats, cause I am a coming to him." He gasped. "I am a coming." He grunted. "To *him*."

Henry laughed. "You hear that, Ralphy?" he yelled back into the dark cave of the shop, "John David's boy sounds like he knows how to work."

"That's cause he's John David's boy, Henry," said a voice from the darkness. "That man don't break for no other man but Jesus Christ, and even then only at the second coming. Will you shut up and help me work this piece out? You know I can't fix'ese by myself."

Henry waved his oil rag at Remmy like the white flag of surrender.

Remmy tried to lift the thing again, but that old hinged claw wouldn't work for him, so he fiddled with it, working all the angles til he found the divots it'd picked into the ice's sides and finally got to going again.

It was easily past noon when he came limping into the store and slammed down his haul right in the middle. Not worth fifty cents. But it sure as shooting wasn't worth a penny less. "Here's your ice," he said. "Here's your tongs," he said. "And here's your change." And he handed over the whole figure. "And you owe me fifty cents."

Mr. Tolliver looked pissed, the giant craters on his acne-scarred face writhing like a pit of snakes. "Don't put that messy thing down'n my good floor, kid. Bring it back here."

Now.

Remmy'd half a mind to give Mr. Tolliver the whatfor about the terms of their deal, even at eight years old, but he was also a kinder young man than most, so he said, "All right."

"Bring it back to this scale, kid."

When Remmy lifted up the ice, Mr. Toliver saw the muddy mess he'd left behind on the floor and said, "Did you set my ice down'n the dirt, boy?"

"It got heavy," Remmy said, "And The Good Lord did not give me the strength of Samson."

Tolliver chuckled. "Even still, I didn't pay you to bring me back cold mud."

"You didn't pay me at all, yet."

"Don't you sass back, I'm keeping a nickel of your pay."

"Yessir," Remmy said and started, tired as he was, to haul the thing to the back of the store.

They got back there and Mr. Tolliver heaved it quickly up onto the scale. "Why, there's almost seven pounds here. That's way, way more than I paid the ice man for. That thing must have weighed ten when you started't the other side of town. How old are you boy?"

"Eight."

"Eight years old and hefting ten pounds of ice all that way. Well, I'll pay you a dollar for your work."

Remmy thought for a moment and said, "How about ninety cents and a pound a ice for myself?"

Mr. Tolliver smiled. "All right then, but it'll be the dirty side." So Mr. Tolliver started chipping and sawing until a pound and a half lay in the tray. He swept it into a bucket. "Bring me back my bucket, okay?"

"Course I will."

Well he took that bucket, which weighed like nothing, as little as it was and with as little ice as there was down in the bottom, but he lugged it around the corner and down the street and left again to his house and took off a towel. His dad's out in the field somewhere and his mother's asking him what exactly he's doing.

Well he said, "Nothing much, Momma." And took the towel and dusted off the dirt from every piece he could see just like the diamond thieves in the comic books.

Now from where he sat, he knew that Mr. Tolliver sold ice cubes for two cents apiece and people never really needed more than two piece in their drinks, and most of Remmy's buyers were his buddies coming back from the fields or from a fishing hole or from playing with the legendary dead rat and the string to swing it on. So he went right to work with his mother's ice pick and got a whole bucket full of ice. Well there had to be a couple hundred pieces in there at least, but it was hot and he'd have to hurry, goddammit.

"Easy," The Good Lord said to him.

"You're right. And thank you for the patience of Job, Good Lord."

"You are welcome. Be quick about it, but stay civil."

To hurry. He went right out to the bank of the river where his buddies would be fishing and said, "Hey boys! I gotta bucket of ice that's gonna melt if I don't get rid of it cheap! It's a firesale right here, boys!"

"How much?" asked Jack Bandy.

"How about a penny for two pieces?"

I don't know how fast word spreads in your county, but in Marion County, Little Egypt, Illinois, it didn't take long for word to get around to all of the other kids that Wilson Remus Broganer's selling twice the ice at half the price (Remmy's words) and in under fifteen minutes he'd sold the whole bucket, even the stuff toward the bottom that'd been muddy, for a dollar'n a half cause some of the kids couldn't make or didn't want change, happy as they were to have it.

Well it was just about as much effort for him to walk home with those pockets sagging full of change as it was to get the ice. He didn't want just to have a place like Camelot and be an old King Arthur. He wanted to make it easy for people to come and stay with him like Robin Hood. He wanted to be Robin Hood inside Camelot and he wanted Camelot in the wood and wild. So he took all that money in his sagging pockets along with the ninety from Tolliver's block, and the extra money left over from the strawberries and the Barbed Wire telephone company, and stuffed it all in the old tin can with the words CAMELOT, MY MERRY MEN, AND OUR PRANK scratched into it next to the spool of barbed wire. Less the Daddy John tax, that is.

The money had started to snowball — it was up to three dollars and thirty-five cents — and he wondered how he'd get himself some more and buy his dad a vacation or another farm.

II. Sonata-Allegro

Among the hills a meteorite
Lies huge; and moss has overgrown,
And wind and rain with touches light
Made soft, the contours of the stone…

— Jack Lewis

WILSON REMUS

1944

A BUS LINE RAN from the Brown shoe company, and from there… well two routes went: one from the park to the shoe company and one from the gasoline station to CNI to the west edge of the business district. That's where this little burger joint stood, where Swiftie's gas is now in Bellhamer. Ten cents apiece. Ten burgers for a dollar. Reben's, they called it. He went there and spent a dollar from his CAMELOT, MY MERRY MEN, AND OUR PRANK fund and got ten burgers and gave two to his sister Gwen and gave a couple each to two of his buddies and ate the rest. Four burgers is a lot of food for a nine-year-old and he felt sick.

Down the street sat a comic book shop, so he spent another twenty-five cents on a comic book and then yet another ten cents on candy from the money that he'd saved. In the comic books, he generally liked the cowboys. He never did go for way out there stuff about Mars and spaceships and robots. He said it would never happen, but it did, didn't it? Nah, he went for the cowboys, sometimes the detective ones. But he found one that day named THE GLORIOUS DAYS OF ROBIN HOOD AND HIS MERRY MEN. It was "A Charlton Publication" and it only cost him ten cents. He couldn't resist a good rendition of the ranger that robbed from the rich and gave to the poor. Robbed from the rich English and French kings and gave sometimes even to the poor Scotch-Irish-and-Welsh people like the MacGills and Brodys and Dempseys and Foleys and gave it back to the poor. He was a bit of a jester, Robin Hood. Like St. Francis and *le Jongleurs de Dieu.* Even the cover of the comic book had the shields of the Welsh and Scottish on it, as well as Richard the First and the French Kings with their gilded lilies.

He needed a bow and arrow.

Back in those days, they had these little popup shops, kinda like how outlet malls will do clothes or how some of them master knitters in Brooklyn will do for yarn at the temp stores. Only back in 1944 in Southern Illinois, they sold whatever they could buy in bulk. So if they could get ten tubs of wholesale peanut butter, a bunch of tin toys, and a couple of sleeves of crackers, well they'd divvy those up into sales bins and then sell them all for the same price. The bins were old scrap shivs, the rough wood and the tin tacks you'd get pricked on and splintered carrying bad. Four dollars for a bin.

First time Remmy saw them he said, "I'll take one of them bins for three."

"It's four," this old haggler said.

"I know, but I'm offering three."

"And I'm all out of three dollar sales bins. I'm offering four dollar sales bins."

"Shoot, lady."

So since he didn't have but two dollars and something left in the MERRY MEN fund, he and Gwen went from there to do a little job at that temp store with the sales bins. They said they'd pay them two dollars to take a sales bin to every house. Remmy charged a dollar extra. So he and Gwen split sides of the street and they knocked on doors. Most people were home, and they thought about leaving them bins on the doorsteps where no one'd been home. But they did it. They took the high road and made sure every sales bin got a home. When they came back to the old gal that ran the temp store, the one that'd given them the sales bins to deliver, she asked them where'd they'd gone, and Gwen rattled off a bunch of states: "Ohio, Illinois, Indiana, Missouri, Connecticut—"

"Ain't no Connecticut street," the old haggler said.

Gwen blushed. and said, "Meant College Road."

The haggler eyed her.

Remmy watched Gwen, whose eyes swirled around trying to find purchase on anyone and anything and, unrescued, she continued on with, "Broadway for a little, and Boone and—"

"Daniel Boone. What a man. I would have married a man like Daniel Boone, wouldn't you?"

Gwen hesitated. "I guess, ma'am. I like boys with an adventurous spirit a bit, but I really like men like Thomas Edison."

"Inventors," she said. "And you? Would you marry Daniel Boone?"

"I wouldn't marry a man, ma'am," Remmy said.

"And why not?"

"I'm looking for Guinevere and Pocahontas and Sacagawea and Maid Marian and Mother Mary."

"That's a lot of wives, Remmy, and one of them's a virgin."

Gwen spat. "I doubt you could handle one woman like that, let alone five."

But the old haggler, she said, "Mother Mary's single."

"No ma'am," Remmy said. "She married St. Joseph. They went together and made St. James, the brother of Jesus."

"James was his cousin, you little smart ass."

"Thank you ma'am, and you have a lovely day," Remmy said and he ran off before she could say anything. And around the corner, Gwen caught up and he gave her half. It was hot and they'd walked pretty far and those burgers started rumbling down in his tummy and he threw them all up right in the lot there and Gwen looked sorry. He felt better, even if sweaty, and a dog ran up and started eating it all.

Gwen said, "No! No! Nasty!" And kicked at the dog.

And the dog growled and then bit her on the leg, right above the knee.

Remmy didn't know what to do. They ran to the drug store, which had a penny sale on for the merthiolate. It was this stronger thing to put on cuts that they had back then. Stung a bit. He put it on Gwen and bandaged her up.

They got home and Daddy John said, "What'd you do?"

Remmy started, "Well there was this dog—"

"No, I mean why did you buy merthiolate?" He pointed to the cabinet. "Already have it here."

"Well how was I supposed to know?" Remmy asked.

"Okay fair enough. But why'd you get two?"

"First one was a dollar and the next was a penny," Remmy said.

"It'll last ten years!"

"It was on sale. We saved money."

"You only save money if you save it, boy, not if you spend it. No sale is worth spending money you don't have. You have anything left over from today?"

They had most of Gwen's dollar left, so they gave him that.

Daddy John kept half and gave them half. He turned to Gwen. "Are you okay?"

"Yes, Father."

"Let me know if it starts to hurt, will ya?"

"I will."

"Good girl."

Remmy went back to his can and sat down to add the coins to it, but now it was down to about two dollars and a quarter. That's what he got for spending his money on the burger deal just cause it was there and throwing it up and getting his sister bit cause of it. Wasn't worth the sale, none of it. He could have come home with five dollars total and instead he had less than he started with. He sat down and read about THE GLORIOUS DAYS OF ROBIN HOOD AND HIS MERRY MEN. Those guys in the comic books built a secret fort out in the woods made out of things they'd found. A whole city of people living in the trees by a waterfall, by a living water, all laughing and carrying on. That sounded nice and cheap and a good way to get away with his Daddy John.

Yes. A bow and an arrow and a good hatchet and some really good rope and maybe one of those army pocket lighters and a knife. That'd do it. That and finding his Maid Marian. You could really make a castle pretty cheap in those days, long as you had the strength to save and flirt and a good funny story for gathering your merry men.

WILSON REMUS

1945

On April 30th, 1945 the boy named Pete Taylor started flirting with Remmy's sister and shoving her. Remmy recalled the date because it was old Pete Taylor. Now Remmy loved Pete. Pete Taylor'd been one of his main men on the barbed wire telephone company. Pete Taylor'd been one of the two kids Remmy'd given burgers to during the burger firesale. But it didn't matter who it was, Remmy wouldn't have nobody shoving his sister, specially one that was chasing after her. Pete Taylor wasn't even no inventor like Gwen'd said she'd liked.

So Remmy smacked his friend Pete with his palm heel.

Pete Taylor stood about a head and a half higher than Remmy at this point in the game, so Old Pete picked Remmy clean up off the ground and said, "Cut that out."

"Stay off my sister," Remmy said, and he hammerfisted the crown of Pete's head.

So Pete Taylor toted Remmy for a block. And for the

whole block, he wouldn't quit squirming and punching and even thinking about biting the hand that held him up off the ground, although he didn't, of course.

"Quit that fighting," Pete said. "Quit it."

"I'm saving my sister," Remmy said.

"All right, Remmy, all right already. I won't touch your sister."

"And you will help me defend her honor!"

Pete Taylor laughed. "What do you think you are, defend her honor?" He laughed again.

"Say it!" Remmy shouted, sideways as he was, being toted into his second block by then.

"I'll say it. I'll say it: I'll help you, Remmy."

Remmy went limp. "Good."

Pete let him go.

He belly-flopped on the ground and said, "Oomph."

They'd walked and been carried all the way to a drive through restaurant with one of those soda machines right outside, so Pete Taylor bought him a truce offering. Three old men were sitting out there eating their fries and listening to a radio. Adolf Hitler had put a gun to his head down in his bunker. April Thirtieth, that's why Remmy remembered.

"It'll be the end of the war!" one of them said.

"In Europe," the oldest said. "The Japanese are stubborn. They'll only give up if you set half the island on fire."

"Well then let's do it," said the third.

"Don't be so quick," the oldest said. "Some fires only stop burning once everything's burnt. Old Pandora had a box she opened, gave her chicken pox."

Remmy snorted and soda came out his nose and it burned.

All three of the old men looked at him and Pete Taylor. The oldest man said, "Scat!"

So they got up and walked away and Remmy said, "Pete, Captain America'll be out of a job now."

And Pete Taylor said, "Thought you didn't like Captain America?"

"I don't. Robin Hood'll never be out of a job. Always gonna be rich guys who need robbing cause they stole in the first place from the poor. Always gonna be bad sheriffs need to be made fun of."

"You and your jubilee."

"Jubilee?"

"It's a year the Jews talk about. Year when all the debt everywhere gets forgiven."

"*Forgive us our debts as we forgive our debtors,*" Remmy said. "Yup, sounds like my kind of year."

"Texarco don't seem like the jubilee type, Remmy."

"We'll see."

"Sure. You're a good friend, Rem."

"The best of friends. Forever."

"You got it," Pete Taylor said.

That year Remmy did some more odd jobs — the deliveries, the strawberries, the ice again, but he added a new one. He became a salesman and sold garden seed packets. He bought them in bulk and packaged them in small amounts and sold them for four times what he bought them for. Pete Taylor did the same thing and bought a bb gun, but Remmy couldn't buy a bb gun because he always gave his sister half the money. Until he got that much money back from Daddy John, who'd taxed Gwen this time, he wasn't about to get that other half. So instead he bought himself a bow and arrow that he could hunt with if he wanted to shoot white tail, but secretly because he was now a *longbowman*. Thanks to the garden seed packets, he was one tool closer to Camelot, where'd he'd have those old castles and throw jokes like jesters at all the dumb tyrants.

Gwen worked at JC Penny that year, selling jewelry and perfume and putting up with the advances of dirty old men.

WILSON REMUS

1946

It was with such dirty old men that Remmy found himself sitting when he went along with his Grandad Patrick to listen to the older men tell drinks and sip stories. Especially in winter, when the old guys got down cause the work was slow and the joints was stiff and you got to thinking about the state of the world, war or no war. Sometimes they actually ended up saying things that made them seem like what Remmy called "fart smellers," by which he meant smart fellers. They didn't always talk down about the women in Remmy's life — and it was a good thing too, because he would have gone at them just like he went at Pete Taylor had they said anything about his older sister. But they didn't. Listening to them old men telling drinks and sipping stories always turned up gems like this:

Rooney Rubinkiewicz said, "They will though, Bullhorn. You'll see. They bounced that radar off the moon, didn't they?"

"Yeah," Bullhorn said, "But that's not what they're going to use them rockets for first and mostly, isn't that right Charm March?"

Charm March was so named because his full name was Charles Murphy Marlon Chowder, for his sins. Chowder like the soup. Charm March also happened to be the way he walked. "You're both right, I guess. Rooney's right about space — they'll get out there and start racing at each other there just like here, you'll see. But hell, everybody's gonna have The Bomb now and it's just a matter of time before they all start using them on one another." Charm March had been the one who'd been rebuked by the older man about fires you can't stop the year before when Pete Taylor bought Remmy a bottle of soda, so he'd come to see reason.

"I just want a way off this rock," Byford Bruce said. "It's just one big Alcatraz, Earth." He was telling his drink. And, sipping the story, he said, "Prison."

"That's not so bad," Herman Hellman said. "There was the Battle of Alcatraz back in the Spring."

"If six men's a battle, Hellman," Abe said, "then it should only take ten of us to get off *this* rock."

"It'll take more than that," Rooney said. "All of us together ain't as smart as Howard Hughes, and that asshole's probably still healing from his crash."

"Cheers to that," Grandad Patrick said.

"It'd take a genius like Orville Wright," Rooney said. "That man landed a plane after a *bird strike*."

"Maybe we don't have to go so far," Charm March said. "Maybe we just have to get out from under Texarco. I hear in England they're nationalizing everything. Nationalizing the banks. Nationalizing the mines. Hell we've got banks and mines. If we could nationalize those and the oil—"

"Sound like a damned communist," Bullhorn said. "England can go to hell."

"Easy, Bullhorn," Grandad Patrick said.

They all looked at him.

"Well Old Churchhill was just here this Spring, wasn't he?"

"West of the river," Abe said.

"Yeah, that's right," Grandad Patrick said. "Fulton, Missouri. Talking at the college there on the evils of the *Iron Curtain.* England can't be all that bad."

"Some people forget we fought the Redcoats," Bullhorn said.

Granddad Patrick slammed down his stein. "And some people forget that I'm Irish and have quarrels with the English older than most of your families."

It got quiet.

Then Abe raised his glass. "To Bloody Williamson."

"To Bloody Williamson," they all said and drank. And when Grandad Patrick finished his beer, they all bought him a new one.

Grandad Patrick said, "Don't talk about that too loud. Eight men suffering in prison for that revolt's bad enough. Don't need to add me or my kin to the mix."

They nodded.

Abe whispered, "*To Bloody Williamson.*"

The men giggled like children.

Grandad Patrick sipped his beer again. They sipped theirs.

"Well at least we're not blowing each other up like the Jewish terrorists, eh?" Bullhorn asked.

"You got it," Charm March said. "What're they called? The Air Guns?"

"Who knows," Abe said. "I haven't even heard of this."

"Blowing up Jerusalem because the Brits tried to bring Israel together. Said it right in the paper. Bombs like the anarchists back when I was a boy."

"Should hang them like the Nazis," Grandad Patrick said. "Just string them up in a high school gym."

Remmy sucked in air really quickly when he heard that. The older men looked at him, but he had nothing to say. He just didn't like the idea of ten men — of any man — hanging from a noose from the highest rafters in the high school basketball gym. Hanging like pirates or like some beggar in the French Revolution Miss Witt had taught him about.

"Well?" Rooney said. "Speak up."

"I…" Remmy said. "I think I'd like to go to Santa Claus Land. Spend some time with Da… with my Father John David, time away from work."

They all chuckled at the boy.

"Wouldn't we all, boy," Bullhorn said. Bullhorn was called that because that was his Native American name, but also because he could be loud when he wanted to, and sometimes when he didn't. "Work is hell!"

"The ground is hell," Grandad Patrick said. "Work is heaven."

Rooney raised his glass. "I'll cheer that on."

"Why Santa Claus, Indiana?" Abe asked.

"Cause," Remmy said. "It's the first park made after a theme. It's like Christmas all over a whole big old park."

"Central Park's a theme park," Byford said. "I used to live there."

"You were homeless in Central Park?" Grandad Patrick asked.

Byford slapped him on the back. "Nah, just in Brooklyn. Went to Central Park a few times, though."

"What's the theme?" Remmy asked.

They all turned to Byford.

"You know, I don't know. Rocky."

"So the theme of the park is… park?" Remmy asked.

Byford leveled his pointer finger at Remmy and then whirled it over to Grandad Patrick. "We don't need two of you, now."

Grandad Patrick just smiled and sipped his beer. "I'm going to take you to a movie soon, just for that," he said to Remmy.

"Soon" happened to be a month or more later in January, and they went and saw not *Miracle on 34th Street*, but rather *It's a Wonderful Life*. Remmy ate it up, he did: the story of a down-and-out banker, a bond salesman, getting taken advantage of — him and the whole county — by a rich, greedy old banker. He found himself crying and recalling again that the only reason to save was to bring his people, whoever they happened to be, together again into a paradise. A kingdom. A fort in the forest. A perfect subdivision of the big city. No, not the big city, he hated the big city. Hadn't Atlantis built a city out of the golden base of the deep? Hadn't they carved a garden, called it out to grow upon the living waters? And where did it get them?

You can imagine his sadness when he heard his Grandad

Patrick a few weeks later ranting and hollering about the FBI memo that showed up in the paper. It read:

With regard to the picture 'It's a Wonderful Life,' *[redacted] stated in substance that the film represented rather obvious attempts to discredit bankers by casting Lionel Barrymore as a 'scrooge-type,' so that he would be the most hated man in the picture. This, according to these sources, is a common trick used by Communists. [In] addition, [redacted] stated that, in his opinion, this picture deliberately maligned the upper class, attempting to show the people who had money were mean and despicable characters.*

But that was the thing. As far as Remmy could tell, most of them *were*. That didn't make him a communist in his mind.

That made him someone who wanted to buy some more issues of *The Glorious Days of Robin Hood and His Merry Men.*

WILSON REMUS

1947

LATER THAT YEAR, he started taking Grandad Patrick to play buckkeeper. They went to a little old tavern with swinging doors like in the westerns, and you could look over the top or under them or walk up to them, if you were tall, with your bucket hat poking over the top and your spurs underneath. Not that people made a habit of wearing spurs in Little Egypt, especially in town. But you could've done it if you wanted to, and that's what made it a neat old place.

Cost a quarter for a game for four guys to play. Buckkeeper. You'd get back four nickels if you won on top of getting your deposit back — one for the building and then the nickels all four of you threw into the pot that hand. Pretty fast game, throwing in nickels, throwing down your cards, raking the pot. Stacks and stacks of nickels came out. They didn't start out on the table like with the poker tournaments they put on the T.V. in later years. Men hid their nickels cause they didn't want you to know how much they's willing to lose, but at the end you'd get stacks and stacks of nickels. You could make fifteen cents every game if you were real, real good. And Grandad Patrick was real, real good.

Sometimes the men would be quiet, so he'd turn to Remmy and say, "Ask me anything you want," and Remmy would ask and Grandad Patrick would tell him a story about how the Cahokia Mounds had been painted by the hands of giants who were cousins of the Native Americans, or about the time he'd swam across the Mississippi just to punch an old cave monster in the teeth, or how he'd once cut down a hundred-year-oak tree with his pocket knife because he wouldn't make it home in time to beat a snow storm to bed and so he needed the warmth of it when he'd worked the whittling wood into a fire.

But this day, Remmy asked Grandad Patrick how he'd been born.

"Me?" Grandad Patrick asked. "I was the child of a leprechaun and—"

"No." Remmy giggled. "Me."

"Oh!" Grandad Patrick said. "Well we carried Midge — your mammy — to the hospital in the back of a horse and wagon, and Midge carried you within her. Midge had dressed in her best Easter dress because she wanted to look good for the doctors before they saw her naked, swollen body, I guess, I don't know. John David — your pappy — didn't want his boy born in a barn or acting like he was born in a barn, for that matter."

"Daddy John was born in a barn?" Remmy asked.

"Daddy John never treated you like a horse's ass?"

The men at the table snorted. Remmy smiled.

"Well," said Grandad Patrick, "Hospital's thirty miles away, so it took the better part of an hour and a half to get her there, and the bumpy road did not help her labor pains. Midge felt them come on strong.

"'John?' she asked your pappy.

"'It's okay, dear,' your pappy said. 'We're almost there.' He snapped the reigns.

"'John!'

"'I'll get you there. I'll get you.'

"'John, I'm having the baby now. Right now. He's coming now, John David, and I'm going to have him right now.'

"'No you're not.'

"But she had you anyways alone in the back of the wagon," Grandad Patrick said.

"The same one when the thing happened with the milk jugs?" Remmy asked.

Grandad Patrick searched the boy's features, seeing him anew and aright. "Was that a hard day for you, boy?"

Remmy wasn't going to talk about that. "It was that wagon?"

Grandad Patrick hesitated. "Yeah, it was that one. And Midge, she lifted up you and smacked you and you cried and she passed out with her dress stained the way the sheets had been stained after your parent's consummation."

"What's a consummation?" Remmy asked.

The men at the table chuckled.

"Big word for love making after a wedding," Bullhorn said.

"Losing yer virginity," Charm March said.

"Anyways," Grandad Patrick said, "John turned around and saw your mammy and said, 'Oh.' He slowed the horses and pulled them over. He tended to his wife to make sure she wasn't bleeding to death or nothing."

It was, in some ways, too late for young Remmy the day he was born. The water Midge had been drinking had lead runoff in it from the oil fields, lead that had been imported from Galena, Kansas and the Joplin, Missouri mines for use in tools and supplies and paint and things. We've known that for a while now. But John David couldn't know that — he only knew and cared that his wife wasn't bleeding. And Grandad Patrick couldn't know it any more than John.

But genetically, Remmy had been born with the highest IQ in hundreds of waves of Broganers and Donders and McGills and Dempseys and many of their forefathers. But even with such good luck, bad choices had neutered the mind of the boy. The lead had dampened that bloodline to make him something like normal in every way but his storytelling mind and his angular mind — the mind that would soon blossom when trigonometry was made known to it. Didn't keep him from reading *The Little Prince* and *Rabbit Hill* and *The Quaint and Curious Quest of Johnny Longfoot* and *The Little White Horse* and *Narnia* and *The Hobbit* that year in 1947, all of which gave him great big images and ideas for handling larger-than-life pains like war. Anyone who tells you reading's escapism is a fool, for one, and for another doesn't remember that all prisoners of war want to escape. John David couldn't know any of this just glancing at his newborn son and Grandad Patrick couldn't know a lick of it telling the story, but there it was.

"So he turned the wagon around and headed home,

stopping at some diner along the way to treat himself to some pickled pig's feet."

"I hate pickled pig's feet," Remmy said.

"That's why." Grandad Patrick played his card. "They named you Wilson Remus Broganer."

"Why?" Remmy asked.

"John David thought Remus sounded like a black man's name. Midge said that's why she liked it. John David asked her if she liked black men and Midge said she loved John David and had forsaken all others, so that was that. But she still liked black men and their culture as such, and John David thought that sounded strange and even heretical to say that and be married still, but he did not question his wife — your mammy — for she and not he had delivered the baby he'd put in her." He played his card.

Patrick didn't know it, but it was those two bloodlines that warred in Rem.

Now about that ancestry:

Around the time of the French and Indian War, a Nuu-chah-nulth man named Wickaninnish moved down from Canada and fell in love with one of the Donder girls — Beth's ancestors — who had been raised as a second wave immigrant. "Wickaninnish" means "No one in the canoe in front of him" or "No one to row with" or "Alone in the boat." Certainly he felt that way — paddling all on his own through life. Even with the Donder girl, who grew cold and distant later in the unlikely union. Wickaninnish found himself wishing he simply had a traveling companion. He was a well-liked man, Wick, but being well-liked in those days was not enough to carry a Native American through to the higher side of society.

Remmy had the rest of the stock that came with the family and the union of John David and Midge, of course — the Donders and the Broganers and so on — but it was the very spittin' image of the Nuu-chah-nulth explorer named Wick that came forward in his features and his nature: wild at heart and joyful in the folk tale, but utterly alone in the boat. The only difference was abundance of chest hair. Remmy came out with a rug already growing. Barechestedness would go to his grandson.

That's me.

Grandad Patrick pocketed fifteen more cents and said, "John David came back out after eating his pickled pig's feet with a hamburger for Midge. Midge put it down in three bites and asked for another. She looked something like a monster from one of those old penny dreadfuls, devouring her food and beautiful on top and something like carnage below, staining her cream dress crimson."

Remmy's eyes were wide and staring at the middle of the table, his mind racing.

"See there, Pat," Bullhorn said, "you went and contaminated the boy's mind."

"No sense in lying to him."

"He'll end up in prison," Bullhorn said, "he keeps acting like you."

"Well shit," Grandad Patrick said, "all good Dempseys do jail time at one point or another. Have two brothers. One's a Pentecostal preacher that heals people and handles snakes. Would've been better off in prison. The other one had a friend who doubted his wife's honor, so he gave him both barrels of a shotgun right to the belly. Murdered him right there on his front porch."

"He do jail time?" Bullhorn asked.

"Ten years."

"Ten years *for murder*?"

"Sometimes murder's justified," Grandad Patrick said.

"Listen to you," Bullhorn said. "Just because eight men got off Scot-free in Bloody Williamson—"

"For one, I ain't no Scot. For another, that was a war of defense, which is the only kind of war," Grandad Patrick said.

Bullhorn eyed him. "You killed all of them."

"That's not how it went. That's not why the death penalty was invented."

"They had the death penalty in the Garden of Eden."

"Yeah, but difference is Jews can read and we can't. Ten years is a nice round number, murder or no. Man can change a lot in ten years. Think of Laban and Jacob. Joseph and Potiphar. Every President we ever had."

Grandad Patrick always made enough money to buy lunch and gas to take Remmy to Carlyle. Remmy drove even though he wasn't supposed to. Grandad Patrick would take him along and tell him some story about the Loch Ness Monster or Paul Bunyan and then, loaded up with Bullhorn's and Charm March's and Abe's money, they'd go to Carlyle. That was Remmy's payment: to go down to the store in Carlyle to see the girls down there. It was just a place where the kids in the county hung out. Sometimes they'd wait to get food until they were back in Bellhammer and that time Remmy asked for Reben's.

"Thought that made you sick," Grandad Patrick said.

"Only years and years ago cause I took advantage of their bulk sale," Remmy said, "and ate them all."

"Well shit, Remmy, ten times a dime is a dollar. That's no bulk sale, that's a con job."

Remmy felt worse than ever for wasting his Merry Men fund on con-job-burger-puke.

Land in Little Egypt rolls on green and black and blue like a jolly green giant that got beat up in a fight and passed out where he fell. The land goes on like that smelling of sweet corn and salty soy, of stale alfalfa and black soil, of the memory of orchards turned cider and that faint scent of lathered horse herds that might one day return, rolling little hills where the giant's nipples or belly might have gone, a land you could see forever if it wasn't for the trees. Flat as Texas and luscious as the Amazon. They used to say that back when the pirates first came and stole the land from the natives, a squirrel could have gone from Maine to the Mississippi River delta in Louisiana without touching the ground, what for the trees. If that ever was true, the last bit of land that held onto that kind of richness and those kind of trees was there in Southern Illinois, there in Little Egypt.

That's before it got raped by oil and coal and giant global farming companies, of course, bled dry by vampires from some other place that always take and then leave and never give nothing back.

But those copses of trees and creeks and little rivers are the kind of thing they'd ride through in some old classic car, burning poison into the air to get to an old storefront with hand lettered windows, the kind you might only see now in Brooklyn.

"Now, I'll buy lunch," Grandad Patrick would always say, and he'd buy it with Bullhorn and Charm March's and

Abe's money. They'd fix them a hamburger. "I don't want any beer and Remmy don't need any beer."

Remmy never got the guts to disagree, but until his dying day he was pretty damn certain that back on that evening he'd needed every beer.

III. Scherzo

. . . Thus easily can Earth digest
A cinder of sidereal fire,
And make her translunary guest
The native of an English shire. . .

— *Jack Lewis*

WILSON REMUS

1948

GRANDAD PATRICK WAS a mean old ornery old cuss, but his mean and ornery came out so that he always did something funny for his friends and his grandson like Remmy would one day do for me. Grandad Patrick was one of the few people that could make Remmy relax in the days following their move from the farm. For a while, long before Remmy or Momma Midge came along, Grandad Patrick lived out at Patoka in a basement under the house, which had a root cellar to keep stuff for the winter — you know, root crops like beets and potatoes and onions and carrots. You replant them, see, down in the root cellar, and they were good for months.

Well Remmy'd heard that one time his Grandmammy'd shouted up at his father, "John David get in here: your dad's acting up!" And Daddy John'd gone down in the root cellar as a kid to see Grandad Patrick with a massive orange carrot sticking out his fly like his you-know-what. Grandad

Patrick said to Daddy John, "Got a big enough sample size after picking these suckers all day long to know for sure mine's longer'n average." Which probably wasn't true and all, being Irish and all. Many of us Broganers — specially the ones in the Dempsey and MacGill bloodlines – are endowed with many things, but that is not one of them. Oh Daddy John was rolling at that carrot in the fly gag, Remmy heard. That's the kind of ornery old cuss Grandad Patrick was. Remmy would have loved to just spend the rest of his life with the man: Grandad Patrick knew how to rest and have a good time with the ones he loved.

Grandad Patrick wrote a lot of letters in those days. He said you needed to know how to write good letters in case you were ever serving as an ambassador or serving a prison sentence and specially if you were serving a prison sentence as an ambassador. Stationery ran him 10¢ a package, and he'd let Remmy help sometimes to practice his letters or at least he'd read them out loud while Remmy listened and Remmy would just laugh and laugh. He wrote Dodge Motors:

Dear Sirs,

After trying your labumba of a car, I can say the only thing you Dodged is my next purchase.

And Dodge didn't care none. They wrote Grandad Patrick right back:

Oh good. We were worried our reflexes were slowing.

Grandad Patrick bought a battery charger put out by

General Electric just so he could return it and the warranty with a letter that said: *Sorry, looking for a more specific charge.*

But the best of all came from the letter he wrote Phillip Morris. He used to smoke Lucky Strike, and he switched for just one day to Marlboro Reds. After tasting them, he got pissed, and his Irish blood kicked in, and he got his letter writing stuff out and said, "Remmy get over here and watch at how it's done by a professional. This is how you get what you want out of a big old mean company like Phillip Morris."

Dear Sirs,

I bought your cigarettes on the 14th of June, this year of our Good Lord, 1948. I wanted smoke and tobacco. You did not give me smoke and tobacco. You gave me hellfire and brimstones. Your cigarettes are without a doubt the worst in the country, this great nation of United States. Your cigarettes are made out of horse shit and alfalfa.

With sincerity and great derision,

Patrick Dempsey

"You wait and see," Grandad Patrick said.

Remmy, who was every bit of thirteen by this point, asked, "What are they going to do, Grandad?"

"They'll send me back six cartons just to shut me up. You watch. This is how you work them big companies."

"Like The Southern Illinois Coal Company?"

"And the Illinois Chamber of Commerce and The Chicago Tribune and..." Grandad Patrick eyed Remmy. "What's that you said?"

For as long as he could remember, Bloody Williamson had been a secret phrase with hidden meaning, like some code between old soldiers. That imbalance plagued him. "Can you tell me about Bloody Williamson now? Whenever we go to play buckkeeper, men buy you beer cause of what you did and I've never heard the story."

"When you're older."

"I'm thirteen. I'm older than most bar mitzvah boys over in Germantown that get to read that sexy book in the Bible."

Patrick leaned back and lit one of his Marlboros. "Damn that really is godawful." He put it out. "Maybe I don't want another pack from them after all."

Remmy busied himself for a few weeks, waiting for the return letter from Philip Morris. He detassled some corn, cut up his hands doing it, and then he used a grinder on some old shelves for this old man's tool shed, got all the rust off of them old steel shelves. He didn't have to wait long, though.

"Remmy! Letter!" Grandad Patrick yelled at him one day through the front door of the mechanic shop Remmy was working inside. Remmy came out and looked over Grandad Patrick's shoulder as the old man opened the letter.

Dear Mr. Dempsey,

You tried a cigarette we have made and accused us of filling it with horse shit and alfalfa. We naturally resent this remark: there's not a damn bit of alfalfa in it.

Sincerely,

Philip Morris

They both cackled so loud that the mechanic kicked Remmy out for the day. So they spent the rest of the day at Grandad Patrick's, rolling horse shit cigarettes and putting them in a Philip Morris cigarette box, just in case they ever needed one. "Keep this in that humidor I gave you, put them back every night and they'll keep." He patted the rolled shit gently. "They'll keep."

Remmy's hands smelled horrible for a day and a half and his mom used steel wool just to get the smell out and he bled and bled. Turned out horse shit's pretty easy to wash off, he'd just gotten some on his clothes and hadn't noticed. But Remmy didn't care. He was giggling because he'd swiped all of Grandad Patrick's letters, and he was feeling them in his shorts pockets.

Granddad Patrick picked on his friends too. He thought it was funny, but it wasn't that funny for whoever was on the receiving end of the joke.

For instance, Grandad Patrick was working on a farm in the country. This poor old man had milk and everything — he'd done dairy farming for twenty-five years. The old boy was doing the milking, and the cow got to kicking. Kicked his bucket over. Well Grandad Patrick was on the other side of that barn, you see. He'd found a hole in the barn wall where an old knot used to hide in the wood. And he'd gone out there to that hole and shoved a big old reed through it to poke at the cows. So after one kicked over the full milk bucket this twenty-five-year tenured farmer'd been milking, old Grandad Patrick came around inside and said, "George, what's the matter?"

"This cow never did that before," George said. "I've worked her for years."

"You're pinching!" Grandad went under there and milked the dairy cow fine. "You see?"

"I must of been," George said.

"Well it's all right," Grandad said, "we all have our bad days," and he walked off around the barn, went back to the hole with the reed, waited till George went back to it, and he poked that cow with the stick again.

The cow kicked over another half-full bucket of milk.

"God dammit!" George said.

Oh Grandad did that about three or four more times till he came back around the barn. "I heard you cussing, George."

"I milked that cow for years and she never acted thataways. I just don't get that, acting thataways."

"You gotta treat 'em like the nipples of your wife." Grandad worked with the cow again, slow and gentle like.

"Thanks, Pat, you're a real friend," George said.

Remmy thought this was hilarious when Grandad Patrick told him what he'd done, but Remmy also thought it best to tell George that night what'd happened, just so that it wouldn't *keep on* happening, you know? It was always that kind of thing with Grandad. He didn't always know when to turn loose.

There was this railroad you'd cross when you're going over to the college and Shadduck. Well back in the day, that railroad came through Boulder on the lake. This was before it was a lake, of course, that came later. It was just Boulder, Illinois back then. Anyways, Grandad had a friend named Old Man Rooney — guy who drove the flatbed to school – who was a widower. Poor Rooney.

"I just don't know what to do to get a wife, Pat,"

Rooney said one day when they were sipping lemonade over at Rooney's house.

"Buy one from Sears and Roebuck," Grandad said.

"Oh quit it, will you?" Rooney said. "I'm serious."

"Well me too, Rooney. What do you think I just make shit like this up? I ain't that smart, you know me."

"Yeah..."

"Well I send companies letters all the time."

"You do?"

"Well sure. I just wrote Philip Morris the other day!"

"You did?"

He told about how there wasn't a damn bit of alfalfa in Phillip Morris cigarettes, and Rooney had a good laugh.

"Send them a letter," Grandad Patrick said.

"Well how do I know what she'll look like?" Rooney asked.

"What do you think that bra and panty section is for?" Grandad asked.

"Bra and panties?"

"Not only," Grandad Patrick said. "It's for ordering you a bride if you're lonely and your first one died in childbirth just like yours did."

"Really?"

"Oh sure. Write them a letter. Tell them you want E4 or whatever and they'll send you one in a bag."

"A bag with a wife in it, I'll be damned." Rooney got out a catalogue and looked. "Is it the price on the column or is it—"

"Oh no, no, no, they can't show the price for something so precious. It's tip-based."

"What do I tip them?"

"Well don't pay too much until they get here and you're satisfied," Grandad said.

So Rooney filled out his letter and put some cash in an envelope and sent it off to the post office. Grandad knew the post office guy — it was Hellman — and he let him in on the joke snatched out Rooney's letter before it could make it to Sears and Roebuck. He opened her up, pocketed the cash, and then read the catalog number.

Then they got a wig that looked pretty close to that catalog number and sent a letter back on some really nice stationary telling Rooney what day she was gonna arrive. Well that train came from Shadduck and stopped there in Boulder and Grandad Patrick was waiting there with Rooney as one of the teachers from the college came out escorting a young man strapped up in tow chains and dressed up like a young woman in that wig. He'd bribed the boy with extra credit. Boy wanted to go to Broadway.

"Oh boy, oh boy," Rooney said. He couldn't see too good, and plus they did a pretty good job with the makeup. Rooney got the man dressed as his bride into his buggy and he loaded her luggage and then loaded his own and while he was back there loading, he-dressed-as-she whipped the reins and ran off with his buggy and luggage too.

Rooney stood shocked.

"Whelp, that's the way it goes sometimes, Rooney," Grandad Patrick said.

"But I didn't even tip her," Rooney said.

"Guess she just wanted your underwear," Grandad said. "You'd think she'd know how to order men's underwear from a catalog. It's just a page away from her."

"But I don't get it," Rooney said. "She was a nice-looking

lady. Maybe the prettiest woman I ever did see. She was gonna be a nice wife."

"Sorry Rooney."

He never did tell Rooney the prettiest woman Rooney ever saw was actually a man.

Of course that's a terrible thing to do to a widower, and even worse when you find out that people actually do order brides who are kidnapped and turn them into nothing more than slaves. Such things shouldn't be. But even still, that joke's exactly how it happened with Grandad. It was always something.

But it went both ways. You can't teach your grandson Remmy all of that, can't tell all them stories, can't activate that bloodline in a boy and think it won't come back to bite you.

Two or three of them — Remmy and his friends — drove out through the country in his Grandad's car. "Man," Remmy said. "I gotta go to the bathroom." And he had to do more than just pee. It was enormous. These '41 Chevy's had a bumper over the top that curved up like a bad lip job, big old bumper. Well Remmy did his business by the side of the road and then picked this turd up and put it on the bumper of his Grandad's car.

At the time, Grandad worked at a dairyman over between Odin and Salem and Bellhammer. He didn't know anything about it. He went to work and the next morning the milk inspector came by and said, "Who the hell shit on your car's big old bumper?!"

Grandad inhaled his cigarette slow and without flinching said, "I know who did it." He exhaled. "What I don't know is how in the hell he ever got up there."

He left it on the bumper until he figured it out. Which also gave him cause to brag on his grandson to half the town, so to speak.

Reminded him of that joke about the guy who went into his hotel with a shared bathroom. He had a pain, but the bathroom was occupied by the person in the other room. He had to go so bad, he crapped in his sock. He opened the window, spun it around and around to fling it outside like King David's sling, but the sock had a hole in it, so he splattered that crap all over the room on account of that hole in his sock and on account of the flinging. So he called the bellhop. "I'll give you $5 to clean it up." And he went to the pool.

Later the bellhop called him at the pool and said, "Keep your money. I'll give you $10 to tell me what shape your ass was in when you did it."

WILSON REMUS

1950

REMMY STILL COULDN'T afford his castle, but he could afford a good time once or twice a year. Tried to get Daddy John to come, but he was always working his love out on the oil fields. Always wanting Remmy to join him in that working love, whistling while they worked and doing things together. Remmy would much prefer to sit at the end of a dock and sip some sweet tea with his daddy and talk some. Remmy knew he'd join his dad's working love for good one day, but until then he was honkey tonking over at the county fair, getting two orders of ice cream.

They all went out to this grove where they baked a pie every year along with all the other church families, went out there to hang out and eat fried chicken and make blackbird pie. It was a bunch of boys and men standing neath these high-boughed evergreens with the ground bleached white from birdshit, shooting up into the cacophony, and four and twenty blackbirds would fall down every time, and they'd make a pie. Just ka-pow and birds would rain down, just tons of them, so loud they couldn't even hear the guns go off. They'd *sing a song of sixpence, a pocket full of rye. Four and twenty blackbirds, baked in a pie.* Remmy was using this old double-barreled shotgun that kept going off with both barrels, like it or not. Killed tons and tons of birds.

A couple of the ladies and a couple of their men got mad at all of the bird killing finally, and combined with this ruthless old calico cat that tore into those evergreens one day, they lobbied the Illinois senate, calling and writing, that declared that cats roaming wild was a public nuisance. It passed the Illinois senate. And it passed the Illinois house. And it landed on Adlai Stevenson's desk, and old Stevenson vetoed the bill and made himself a public service announcement:

It is in the nature of cats to do a certain amount of unescorted roaming. The problem of cat versus bird is as old as time. If we attempt to solve it by legislation who knows but what we may be called upon to take sides as well in the age old problem of dog versus cat, bird versus bird, or even bird versus worm. In my opinion, the State of Illinois and its local governing bodies already have enough to do without trying to control feline delinquency. For these reasons, and not because I love birds the less or cats the more, I veto and withhold my approval from Senate Bill No. 93.

With the word back from the Governor, the boys and men went back to shooting four and twenty out of the trees. But old Adlai got into trouble for some scam in later days.

Remmy got himself a '37 Chevy two-door with a six-cylinder engine in it, floor shift. Man he went everywhere with it. All the boys would pile in and wanted to go. He bought the car and insurance himself. Called it his trusty steed, even though it wasn't a Jeep. Jeeps couldn't run for shit anyways. And even if they could, he couldn't afford one anyways. Well the boys would jump in his trusty steed and wanted to go.

"Pay for the gas or you ain't going," he'd say. Remmy still had his savings can, only it was bigger then.

"Oh come on Remmy," Pete Taylor would say.

"You come on. Is you a best friend or ain't you?"

"Fair's fair I guess."

They'd get out their quarters and pay him. Sometimes they'd run short on money and wouldn't have enough for C gas or B gas or even A gas. That's when Remmy found out the Chevy would run on kerosene. Couldn't start on it, hell no. But if you got it going hot with a little bit of the C gas, and waited real patient like, you could put it in there and it'd run pretty good. Kerosene, they had a-plenty because they were burning that instead of coal for heaters those days. Coal oil lamps too, with their twisty wicks made of flat rope. Remmy liked to turn the unlit wick up really high and back down and pretend he was a snake charmer. Robin Hood had a snake charmer magician along sometimes in the stories. Arthur had Merlin. Every good Utopia had a magician or a miracle worker. Most comedians travelled with a magic show. He still liked faking with the lamps, even at sixteen. Well anyways, they'd run around that part of Little Egypt in the car that ran on kerosene, putting the C gas in the carburetor to get it to start.

But some days you couldn't even get enough together from the boys to put the C gas in there to start it. One time they happened to be in Boulder, so they went up to

Rooney's house to ask him for some gas, but Rooney was asleep, so they took a hose and a can and stuck that hose down the filler pipe of his Studebaker Commander, which would only run on A gas, and sucked and sucked to get it primed so they could siphon the gas out and into the can. Remmy was giggling, thinking about the boy dressed up like a woman and chained. He hadn't seen his Grandad Patrick much that year cause of the work up north, but he thought about him then and all of those stories came back and he was giggling even as he was sucking and sucking at the hosepipe. And mid-chuckle that A gas came on strong and he choked on it and it burned every bit of lung it touched.

He doubled over, wheezing, and wondered about the state of coal miners in Williamson County that lead to the Blood event. He wondered if Grandad Patrick would ever tell him.

His boys laughed at him as the gas poured into that tin can. He switched between laughing and crying out from the pain, laughing and howling. And he couldn't get the gas stopped either, so once the can was full, he just poured a good bit back into Old Rooney's tank so the poor old widower could get back to the station.

They had three refineries out there in the oil fields. They had this dripped gas. Guess they called it that cause it was the drippings off of whatever the pumps was cooking, hell who knows? People were running cars on it, sort of. Smelled like a volcano when it farts, like the tramp who took The Devil to dinner, burning things shouldn't be burnt.

Remmy couldn't use it because their Daddy John didn't have Texarco on his side, didn't work them enough like some of the families. Perk of the job, I guess, that rank old gas. But if they had it, their car would stink something terrible. His buddy's dad had it, though, and he was working for a gas station. The owner there told him, "You're good help, but you've got to park that car over there, because I'm not having it smelling of that stinking gas."

They drove that Chevy a lot of places. Three or four boys. Girls in Bellhammer. Girls in Salem. Girls in Odin. Pickneyville. Running around one Saturday, something broke and it shut down. So there they are in Pickneyville. It was on a Sunday, not Saturday. So Remmy made a call and Daddy John headed that way. They had an hour fore he'd get there.

Had a baseball bat in the trunk. Remmy kept it in there in case things ever got nasty, cause you never knew. Even at five-foot-nine, he still packed the punch of a six-foot-four-Broganer, a strong brew of Scandinavian strength and Irish rage. Things sure had changed since he'd tried to beat up Pete Taylor for making eyes at his sister. Remmy'd learned the way of the manful art.

Well they'd broken down there in Pickneyville next to the apple field where you did your pickneying. So all three or four of them went out and picked apples, and he got out that bat, and they hit bushels over and again. They creamed those things and smelled of applesauce by the time Daddy John got there with the other car.

Daddy John shook his head and smiled, even though he looked miffed. Remmy got in his car after the chain was hooked up to the other one and put her in neutral. Remmy was happy to have his Daddy John along, even if it was just for working on something together. Driving a car being pulled's a whole lot different compared to driving a car that's pulling its own weight. A lazy car can overcorrect real quick, jitterbugging back and forth across the road. Remmy made that mistake over and over again. Daddy John would steer and Remmy'd overcorrect and Daddy John would steer the other way and Remmy'd overcorrect the other way. He tried, but it was almost like no one was in the canoe in front of him.

Finally Daddy John stopped the car in front. Remmy almost rode up his dad's bumper, but got her stopped.

One of the boys in the pulled car said, "Oh shit."

Daddy John came round. "Get out. Let Little Pete drive. You don't have the know-how with this, Remmy. Get up in the front with me."

Remmy felt the heat in his cheeks and watched as Pete mouthed *sorry*. Remmy went up in the car and his dad moved the seat and made him sit in the back, as if Daddy John were his cabby and left all his friends in the back car.

And he sat there in the back, wondering again where his real dad went, just like in the milk cart when he'd been six years old. Only this time he didn't even have any Lincoln Logs to keep him company.

WILSON REMUS

1951

"A FOR EFFORT," MRS. Elastic said and handed him his card and winked at him. As a senior in high school, Remmy did much better at getting the winks he wanted. He already had all of the credits he needed to graduate. Took what he had to take. Nothing more. What was he gonna do with shorthand? Nobody was gonna write like that in the future. No physics in his senior year. Don't need that. Remmy wanted to be either a carpenter or a CPA or perhaps a real estate broker. Trigonometry looked like it might actually be useful.

Spent a lot of time in the office. As much time in the office as possible, really, detention and whatnot. For his senior prank, he led three hogs up onto the third floor of the high school earlier than the 4-H club showed up at the start of the school day and he'd painted a single number on each of those three hogs in bright blue paint: #1, #2, and #4. Took eight hours for them to get out the three little pigs and clean up the shit. Then all of the teachers stayed up till midnight trying to find hog #3. Yeah Remmy did everything he could to get into that office and stay there.

Went there because he knew the girl working in there. She was just fun to talk to, so he got to talking and the school powers that be had to come to the office and get him because he was there and take him where he did not want to go.

Typically that meant a work site with wood and saws and oil.

Remmy already spent every Saturday with Daddy John, doing working love some place. Insurance — if it ran short for the car, he went down to Grandad Patrick's, who'd loan him fifty dollars and Remmy'd pay it back however he could. See, Grandad Patrick had a little money stashed away in a money sack with a hand painted sign on the front that said IN CASE OF BLOODY WILLIAMSON, and the money sack had some good stuff down in the bottom of it. He was good for more than jokes, for sure. He made life livable every way he could. Sometimes fighting for it with his hands and blood and teeth.

Down the street half a block by the railroad siding, Remmy hauled in limestone for farmers to put on their crops. When they got done, they didn't sweep it out. Grandad Patrick would sweep it out with Remmy, and they'd get quite a pile, and they'd load the wagon and wheelbarrow full. They'd take all of it over and put it on everybody's land, the limestone. That's how he got a coal car — there'd be pieces of coal in there at the bottom of the limestone bed beneath the white limestone, and they'd sift through it as they spread it out for folks. It was a truck with a load of free limestone that had free coal in the bottom, but they called it the coal car in the journals.

Remmy'd get in the car and toss lime and unearth the black rocks hiding down on bottom of all that limestone and they'd get a little coal to burn. Rough times. Wasn't anything flowing free like it later got.

He started reading books on carpentry, and Daddy John started taking him along to different scaffolding or house building jobs here and there. He'd read a novel and then read a book on building the studs of a house, and then he'd go to class and repeat. He learned about what tools they used to build the pyramid and what angles they use in the barbicans of castles and how that works for the gable of a big house. Somewhere in there, he learned about how a trebuchet works — a word that he pronounced "treb-you-chet" for the same reason his neighbors pronounced "Cairo" Illinois "Cay-ro" like Caro syrup. That trebuchet had circles and squares and triangles all on it — its base was isosceles and its arm was ram-rod straight and it had a little sling in it just like the joke about the guy with the sock. And the counterweight was on a pivot so that, as the arm dropped, it would hue to a straighter line for a little extra oomph. He worked those trebuchet designs over and over and drew so many different versions because it had so many trigonometry thoughts built into it. That was the only thing in the whole trigonometry book he didn't really apply in the real world, those trebuchet designs, first on account of it being useless in Southern Illinois and second on account of it being something out of the old comic books he'd read as a boy. But most of the other stuff he was able to use on the different building projects Daddy John took him to. It wasn't a vacation with Daddy John. And it wasn't paradise or the promised land either.

But it was a workable love because it was a love of work. So they worked with that and Remmy saved from the building, borrowing from Grandad Patrick as little as he could.

After his first few real jobs, he tried to buy Daddy John a vacation for just the two of them: just time and one another and maybe fishing or something fun. They were mixing concrete to pour, mixing it in shovels and hoes and wheelbarrows just like Frank Lloyd Wright did when he built Fallingwater. A house on top of living water. Remmy brought up the vacation and using his money to pay for his dad.

"You know I don't do vacations no more," Daddy John said.

"Never again?" Remmy said. "Not with me at least?"

"Not until your momma's dead and gone," Daddy John said. He had a point, but Remmy hated that point.

All Remmy recalled from growing up was that Mother couldn't travel cause she got a sick headache every week. Every dadgomn week, there was no end to it. She ate bread and got a headache. She ate cookies and got a headache. She'd sit too long and get a headache cause she wasn't standing and then would stand up and get a headache cause her feet hurt.

Well they went down into Kentucky once. They never went on many trips — he or his sister or Daddy John along with them — because Mother would get sick. Remmy figured he was four at the time. Well, on this trip, the river roads and the winding mountain passes made her sicker and she screamed, "John David, pull over, pull over, pull over!"

And they pulled over. And she got out and leaned over her knees.

Nothing happened.

"I've never seen anything like it," Remmy said.

John David wheeled around on him and his sister and spat out words while their mother bent over outside. "You listen here boy, don't you make things harder on my wife than they already are, do you get what I'm telling you?"

Remmy winced.

"I know you get me. You leave her alone, you hear?"

He left her alone. And in leaving her alone, felt he'd been left alone himself, exiled from the pack.

Of course, at four, he'd never seen anything like anything. He'd never seen anything like grass. He'd never seen anything like cracked floors. He'd never seen anything like *tall.* That's half the fun of living: seeing something you've never seen before and never would have thought of seeing until the moment you did. Socrates said it, didn't he so? Wonder's learning something.

So in a way it was still new to see a woman suffering all the time like that. And he wondered if the suffering wasn't more her own doing: if she was somehow punishing her own body by way of her mind, letting her worry wither her away, letting herself worry because no one else would worry for her, which really meant that everyone else got to have all the fun, so she had to make them pay her back somehow for her self-martyrdom. Cept it wasn't martyrdom, not really, since her only cause and only belief was for and in herself, and ain't no martyr in Foxe's book that died for himself. That's called suicide. More on that later. Anyway, they cut the trip in Kentucky short, and Daddy John never went on a vacation again.

All Remmy wanted was to go fishing with Daddy John, but mother must not have liked watching him have all of that fun, because she got that lazy hatred about her every time they'd gotten off the farm, this thing that worried her sick so that she got the attention and her husband did not. Such things made it hard to rest in his father's love, Sabbatical or no.

Anyway, that's how it felt to Remmy at fifteen: mother making him pay. Oh she wouldn't cuss at him, but she would shout at the kids for cussing. Mother had no education, so the only way she knew how to argue was shouting or making you feel stupid or like more of a sinner than she was. That was specially stupid because the Bible she thumped told her she was just as evil as the rest of them and just as ready to do good if she grabbed onto the same grace she preached. But she didn't most of the time. Most of the time it's like her brain could only understand a few seconds of messages coming in before being overwhelmed with messages going out. That's why she talked loud when she actually did talk. That's why she talked *at* you and not *with* you. She didn't know how to think for herself, to write for herself, to use her voice to express herself. She didn't know how to listen to the folk on the other side of the conversation who knew how to think, write, speak for themselves. And because of that, she didn't know how to argue the finer points in life or be like people with common sense. And so she felt lousy because she couldn't cuss to express herself and didn't have no other way, so she made them pay for her worrying and fretting and moping about, made them pay for the headaches she worked herself into.

Coming home from Kentucky they went through Arkansas and the cottonfields of Southeast Missouri and crossed the river back into Little Egypt. Father always wanted them to see something along the road, but they never stopped because Mother wanted them all to suffer.

And no matter how many times he asked and how much he saved working those real jobs, Daddy John still said, "You know I don't do vacations anymore."

"Never again?" Remmy said and sounded hurt every damned time.

"Not until your momma is dead and gone," Daddy John would say and mix his concrete batch. "Don't you tell her I told you so."

"Yes, Daddy John."

"Sure do love working with you, though."

"Yes Daddy John," Remmy said. It was good. It wasn't the farm.

WILSON REMUS

1952

FORE WE GET to the main girl Remmy tried to date in 1953, you need to know about cat whipping and kettle pitchering — both of them used to humble fools. That's the point of a good prank, you see: bringing low the mighty. But sometimes a prank backfires on the prankster if he keeps on picking on people that need not be picked on, having fun at people's expense. This time it worked out okay, humbling some old fools with the cat whipping and the kettle pitchering.

First, cat whipping.

This is the kind of trick fools in the country will play on one another and it was the last trick Remmy pulled with old Grandad Patrick. This old oil family had moved in when Remmy was a boy, and they had two sons. One named Tom Johnstone, the other named Jim. Jim's the one told everyone in the one room school bout the planes bombing Pearl Harbor, one whose notebook Remmy went and pissed on. Tom got the blunt end of the cat whipping and Jim Johnstone got the blunt end of the kettle pitchering. Tom's a big old strong boy, you see. Really proud of himself, you see. He goes one night and starts bragging, drunk like, about how big and strong he is and can take everybody in town. His dad's an oil man, what's he got to fear?

Smartaleck carpenters.

Grandad Patrick and Remmy go up to this boy at one of the bars — catfish frying joint on the coast of the lake, what with its colored Christmas lights and top popping soda machine — and they went up to this drunk young man, Tom Johnstone.

"I could lick every one of you right now," Tom said. He always talked like that when drunk, you see. It's what they were waiting for.

"Really?" Grandad Patrick said.

"Sure old man. I bet your bones'd break."

"I bet you're not that strong."

Tom stood up and turned around and walked past his barstool to face Grandad Patrick. He's about two heads taller than old Grandad Patrick. "I'll show you just how if we take it outside, old man."

Grandad Patrick scoffed. "Boy I bet you're so weak and scrawny, you'd get pulled through the lake by a cat."

"Bullshit."

The other men got to chuckling.

"No sir," Grandad said. "What do you think, Rem?"

Remmy scanned him up and down, knowing what the joke was on, and he said, "I bet that cat would pull him clean across to the other side."

"Bullshit," Tom said. "Who's betting? I'll bet all of you bastards." Again, oil money.

Each one of those boys laid down all they had. Hell there had to be seven hundred dollars in there. About seven grand in twenty-twenty moneys.

Young Tom laid down seven one-hundred dollar bills, which nobody ever carried around but him, just like nobody carries around thousand dollar bills in twenty-twenty moneys.

Young Tom laid down seven one-hundred dollar bills, which nobody ever carried around but him, just like nobody carries around thousand dollar bills in twenty-eighteen moneys.

"Here's the rules," Grandad Patrick said. "We'll tie you up to a cat across the lake and when the time comes, we'll whip the cat and say 'go' and you'll go till you or the cat's in the lake."

"That's a deal," he said.

"Got a witness?" Grandad Patrick asked.

"I'll be witness and watch the moneys," Rick the barkeep said. And he did and did.

Well they tied a rope around Old Tom's waist that was every bit of three hundred yards long, and it was dark by that point, but they took the boat and started uncoiling it saying, "We'll flash the light and shout when it's time." Well they went across the lake and took this big old tomcat they'd found out in the trashcans behind the bar and they tied him up. And then every man that made a bet grabbed hold of the rope to whip the cat.

They flashed the light across the lake and shouted, "Go!"

Old Tom commenced to pulling.

Meanwhile, all them scrappy carpenters and farmhands whipped the cat *with the same rope that was tied to Tom*, meaning of course that though they were whipping the cat gently, they were mostly pulling in slack after slack after slack connected to Tom. Well, hell, it's night and Tom can't see for shit so as far as he knows, drunk as he was, he's losing to an old tom cat and gets into that water and drug half across the lake before they turned him loose to swim back to shore.

They bought him drinks at least.

Kettle pitchering's how they dealt with his loudmouth brother, Jim Johnstone. See when someone likes the sound of their own voice a lot more than anyone else does, the good old boys in Little Egypt use a strategy called kettle pitchering. Jim Johnstone was that type. And it basically went something like how boys on the schoolyard will play keep away with a ball hog, only it's for loudmouths. Went like this:

About a week after the cat whipping, Jim Johnstone was trying to tell the story to a bunch of people at the ABC in Bellhammer. He'd been talking for the better half of an hour without anyone cutting in, and the boys had had enough of it by the time Jim's version of the cat whipping story came around. So they set in to kettle pitchering.

Jim said, "So Tom was out at the catfish frying bar—"

"It wasn't the catfish frying bar," Remmy said. "It was O'Lunney's, the Irish pub."

"No," Pete Taylor said, "It wasn't the Irish, it was that German restaurant up the hill."

Grandad Patrick said, "Now, I was there. It was on one

of those lake boat pop up restaurants old Bubba runs off his pontoon."

"Pontoon?" Rooney asked. "I heard old Bubba just had a trolling motor."

"You know," Remmy said, "The other day, I saw a trolling motor for a buck fifty lying on the side of the road."

"Buck fifty?" Pete Taylor asked, "Hell, that's the price of a burger here."

"You boys want another burger?" the waitress asked. She's in on it too, you see.

"Oh sure," they said and ordered one for everybody, even one for Jim Johnstone.

That took a good five minutes getting all those orders and beers for the older ones (and some of the younger ones that looked like older ones, or near enough) and Jim said, "Anyways so they're out there, and Tom got to talking tough like he does and started saying how he could lick every single one of us."

Grandad Patrick asked, "You ever been licked by a man, Rooney?"

"I've never been licked by a woman. Have you, Pete?"

"Now Rooney, you know that's no civilized talk for such a fine dining establishment as a restaurant and bar and grill named after the first prime letters of our alphabet. I will only say, though, that the tongue is a strong muscle."

"You know," Remmy said, "I heard the Mexicans down in Williamson County—"

"BLOODY WILLIAMSON!" they all shouted and bought Grandad Patrick a beer, who grinned an evil grin.

"—use cow tongue on their tacos."

"They call it taco linguistics," Grandad Patrick said. "All Greek to me."

"You know," Remmy said, "the Presbyterian Reverend took Greek in seminary."

And away they went, and Jim Johnstone stewed and violently chomped on his ground beef. He had on a jacket with the Texarco patch, company man that he was. He stared at the waitress as if he'd like to stew and violently chomp on her. He couldn't quite figure out why he couldn't get a word in edgewise. Truth is most men who complain about not getting a word in edgewise are really just loudmouths and spoiled sports who won't shut up about themselves most of the time, so the most dignified way the group can keep them from talking — since shouting WILL YOU SHUT UP? at a man in public's pretty tacky — is kettle pitchering. Keep away for bad storytellers and story hogs.

Tween that and cat whipping, you can keep the tough talkers straight.

For awhile...

Anyways Remmy started trying to date his main girl mainly because Gwen went and got engaged to this good old boy named Ryan, whose ass Remmy once tried to whip and who'd responded by saying, "Well Remmy I figure we can either go outside and whip each other until we're swollen and bloody and mad and concussed or I can buy you more drinks than you've ever been bought and we can just keep the swelling isolated to our livers and we can get red-blooded and happy and pass out and you can call me brother."

"Hey boys!" Remmy said, "I just bought me a brother!"

That made two Merry Men: Pete Taylor and Ryan, the one he'd earned and the one he'd bought.

Remmy forgot late in life that that was his first drink. He'd count a different one as first.

WILSON REMUS

1953

REMMY TOLD PETE Taylor, "You see those two girls?"

"Yeah, they're pretty."

"I'd love to date the one in the box," Remmy said.

"Well let's go get 'em," Pete said.

"Will you go out with me?" Remmy asked her.

"No. Leave me alone," she said.

Every Monday.

See that girl shifted out of the school office and worked over at the ticket booth of the movie theater that later became a regional theater.

"Will you go out with me?" he asked her.

"No, now leave me alone," she said.

Every damn Monday. Like liturgy.

"How's your girl in the box going?" Pete would ask. He'd already dated and dumped his.

Remmy stewed. He stayed home for a week and read

The Caine Mutiny, a book about a nonviolent mutiny on a ship — one involving the law.

Finally after Christmas, Beth went on one date with him. He spent the whole time telling her stories about the family and Grandad Patrick. Well Grandad Patrick kept his shit up until he was seventy-five and Remmy was eighteen years old. Then Grandad Patrick died one day out in an alfalfa field, of all places. Probably shat himself too, like you do when you die, which made Remmy laugh to think about. At the funeral, crying, and at the wake, laughing as everyone shared stories over their beer and whiskey, Remmy wished he had enough money to buy one more day of laughs with the man.

The men stood around and said, "To Bloody Williamson."

And they cheered and drank in Patrick Dempsey's name.

"What happened in Williamson County?" Remmy asked. "He never did tell me." He figured he had a chance with the question because some stories you don't hear until a man's dead and gone. Sometimes that's the only way to get to the whole of a man is to listen to the songs everyone else sang about him.

"Come on and sit down," old Abe said.

They bought Remmy a drink. He came and sat down.

I reckon that's when he sipped his first drink. According to him. He didn't like it much. Tasted bitter.

Rooney started in.[1] "Early in June of the Year of Our

---

1 A whole bunch of the next chapter oughta read like sandpapered block quotes right outa Angle, Paul M., *Bloody Williamson*. 1992. Chapter 1. Tried to think of ways to dress the whole story up, but sometimes — like I said right out the get go — truth's stranger and there ain't no better way of telling it than Angle's way, so make sure you go and check

Lord Nineteen and Twenty-Two, it was the Southern Illinois Coal Company that went and done set up shop with one of them strip mines."

"Whereabouts?" Remmy asked.

"Midtween Herrin and Marion."

"Williamson County," Remmy said.

Abe nodded.

Rooney said, "November in Twenty-One."

"October," Abe said.

"November, dammit, these guys sold their first bit. Regular old dug mine campaign was up through that April. Fifty man crew, wives and kids, you know."

"Old English blood came down," Abe said. "William J. Lester. Came to strip it in Twenty-Two, graduated from one of them big city slicker schools. Harvard."

"Cornell," Bullhorn said.

"It was Columbia," Rooney said.

Bullhorn mumbled, "Pretty sure it was Cornell."

"You kettle pitchering him?" Remmy said.

"Anyways," Rooney said, grabbing the story's reins, "he's one of them smart civil engineer types, did strip mines all over when he could, flipping The Good Lord's earth like it was flapjacks. Lester done went out on his own for Southern Illinois. You know the kind."

"Rich fool promotes himself," Bullhorn said. "Pissed his name in every bit of dirt he saw."

"Course he wants this thing to do well," Rooney said. "Never did it come into his wildest dream of dreams that

---

out his book: I'm about to bullshit you this chapter as bad as any other, but it's all true.
Specially the flies.

the whole thing would fail. Took out a monster loan for tools and supplies and things, so he couldn't spare no slacking off. So, of course, that's when the miners' strike started.

"Strike?" Remmy said.

"You ever work in a coal mine?" Rooney asked.

"No."

"Profit's the only thing them firms like that care about. Don't care about the life and health of us, their workers. Not a bit. We was working in water halfway up our knees. We was in places with air so foul you had to hold your breath going in and if you stayed too long, you'd die a few months later. You ever hear of black lung?"

"Yeah."

"You ever hear of the canary in the coal mine? The one that tells you the air's bad cause it dies first?"

"Yeah."

"We was burning through canaries like they was Christmas quails. Didn't have no laws. No comp for workers — you know the law…?"

"Worker comp," Abe said.

"That one. Buck fifty a day, we made. You ever try to feed a family of five on a buck fifty a day? Union came in Ninety-Eight or Ninety-Nine, I don't remember, your Grandad was about twenty or so and getting into trouble like he always does." Rooney teared up. "Always did."

Several of the other old stoics teared up. One mumbled *Bloody Williamson* into his glass.

Rooney gathered his thoughts. "I's younger. Well they got that law passed and got us raises to seven and even fifteen dollars a day. *Fifteen dollars a day*, you can't imagine. Times tenning our pay. We looked better, smelled better, means our wives liked us better and then made us feel better at night too. Cities grew. We groomed some parks. Saved some of the trees down there and got better trains and cars and tin toys for the kids and things. The whole of Southern Illinois thrived — like we were *sharing* the land just like God intended. It was a promised land. That's why we called it Little Egypt. That and the corn in the storehouses that came later.

"But then Mr. William J. Lester decided to go one step further than the fix he put in with the union. He paid off the powers that be in the union — fixed them real good like a New Yorker or a regular old Chicago Capone — to dig up all of this old coal in that strip mine. And the union said *fine, fine so long as he didn't load or ship none of it.*"

"Let me guess," Remmy said. "He went and shipped it."

"Well that alone was a gift, digging it, brushing it off," Rooney said. "His operator could fill them orders soon as strike ended. So come June that old fool had six hundred hundreds tons of coal dug up: price'd shot through heaven cause of the strike. So he went and shipped that coal. Mid-strike.

"You gotta realize now, Remmy, when he shipped that coal, at the end of the day he was saying to all of Southern Illinois on strike, 'All you worked for in the last age is worthless. It'll rot. It'll go away.' Shipped it and didn't boost no wages. Shipped it and didn't give no benefits. Hell he shipped it and didn't even pay slave wages to the men from Little Egypt that'd worked it before he'd come — stole our whole damn birthright out from under us and didn't pay us a dime in royalties or dividends. Didn't even hire a one of us."

"And that didn't go down too good," Remmy said.

"We didn't want all we'd worked for in a whole age to go away. This was Little Egypt. We was in a promised land, you see."

Bullhorn shouted, "Plus Lester was an asshole! Closed roads all our farmhands'd used going on two decades or more."

"Thirty," Abe said.

"Thirty years. All down the detour we had to make he put hired guns and guards. Irbii's wife—"

"Irbii?" Remmy asked.

"Ireneusz Ritter Blackburn Ishtar-Ironside," Rooney said. "Called him Irbii."

"That's his whole name? Like Charm March?"

Charm March raised his beer.

"Hell I don't know what his name was, I just know his initials made a nickname," Rooney said, "But Irbii's wife was out there by the roadside picking them some fresh berries for a blackberry, raspberry cobbler like you do for the boy scouts, and those old hired guns went and jammed a .32 caliber pistol in her ribs and shouted, 'What the God-damned hell are you doing here? Beat it, and that God-damned quick!'"

"There was another farmer," Abe said. "Every day they stopped him cold and so he just made threats to get legal with their asses."

"State's Lawyer," Bullhorn said.

"State's Lawyer," Rooney said. "And that old guard said, 'You and the State's Attorney can go straight to hell in a handbasket.' So you had a deputy sheriff who had a blow-out tire there by that same road went and flashed his badge, and those coal mine hired guns told him, 'We don't give a damn if you're the President of these United States; you get a move on.' Abe got pulled over by them."

"Yup," Abe said. "He called me a 'God-damned son-of-a-bitching spy.' And the other said, 'Yup, that's just what he is to a T.' And they backhanded me right in the cheekbone."

"Right here," Charm March said, and back handed Abe.

They all laughed.

"Dammit!" Abe said and he swung at Charm March. But they were both old and drunk enough that it didn't amount to much.

"Anyways," Rooney said, "They gut punched Abe with gunstocks and took his small change — called it a *civil...* a... Bullhorn what's it called?"

"*Civil forfeiture.*"

"One of them things. Told him to get a move on. 'You cheep this like the little mine canary you are, and I'll bump you off,' he said to Abe."

"That's right," said Abe between wheezes. He was too old to fight and it was making him laugh, getting his ass handed to him so quickly by a younger, but still old, man.

Rooney was oblivious. "Superintended C.K. McDowell was worse than them hired guns. This asshole? He'd go and say shit like, 'We came down here to work this mine, union or no union. We will work it with blood, if we need to, and you tell all the God-damned union men to stay away if they don't want no trouble. We had our strike and Lester said he'd broken strikes and he'd break ours and we said, 'Come on, then,' cause that was bullshit and everyone and their hound knew it cause this was the only mine this asshole had ever owned no matter how many he'd worked for other white collars. They grew our anger and grew our anger and that was before they brought in a whole bunch of strike breakers from out of state."

"Damned scabs," Abe said from the floor.

"And then they shipped that coal," Bullhorn said.

Rooney said, "Well the Union told us they had our back and called them outlaws, all those scabs and strikebreakers. We held a meeting in the Herrin cemetery — hundred of us from all over the county — and aired how we'd been wronged. I remember it clear as day cause I was sitting on this old steeple gravestone with the name GRAVE on it, which still seems funny to me, being a surplus of the same meaning like it is."

"Redundat," Bullhorn offered.

"I don't care what it's called," Rooney said. "Your Grandad Patrick Dempsey was at the head of all that meeting and debating and hollering, telling us we had to chase them scabs out of the county. So we formed mobs."

"Looted the hardware stores," Abe said.

"Helped ourselves to whatever guns and ammo we wanted," Bullhorn said. "That's back before we all got gun cabinets and closets and trunks like the Holsapple daddy keeps."

"Holsapple likes his guns," Charm March said.

"Don't we all," Abe said.

Rooney said, "The guards heard about this and pulled off three shots and one of those shots landed and killed one of Patrick's cousins."

"My third cousin?"

"Sure, however you count it, and another one tagged another guy I think or… and wounded a third maybe? Oh man, then it was on like a factory at dawn. Men came down from all over Herrin and Marion and wherever else and we had people even from Carbondale and up from Cairo and Metropolis. They came from every which way, all of them whose lives had improved with the unions and all of them sick of them big money companies coming down and stealing our birthright out from under us like they did in Britain, paying us in dirt and leaving us nothing but weeds and dandelion wine. It's damned medieval."

"Feudal," Bullhorn said.

"Slavery all over again," Abe said. "And Illinois never did have slavery so you know that's the closest we got."

"*You work the land, I own it,*" Charm March said.

Rooney said, "Anyways we done shot near five hundred rounds in that first hour out from behind the roadblocks, just like you see now in the wild west shows with over-turned carts and railroad ties and things. I think we used mine carts and a couple of wood stacks and then a bunch of old tin laid out by some cars the owners'd left to rust down.

"Well Old Pat your grandaddy led us all through the night, shooting at the guards and the workmen and any-thing that moved that wasn't us."

"Oh come on," Bullhorn said. "*Everything that moved,* listen to you."

"You don't remember the deer that Abe bagged?" Rooney asked.

"I bagged a deer in the shuffle," Abe said.

"You bagged a deer?" Bullhorn said. "Bullshit."

"I bagged a deer."

"You bagged a deer."

"The seven point on my wall. "

"The seven point on your wall."

"Came back for the rack the next day. Family and some friends ate the meat 'cause you know we didn't have nothing to eat."

Bullhorn watched him for a minute and a half and then said, "Where in my own cotton-picking hell was I?"

"Scalping scabs for all I know."

"That's the Crow, dumbass," Bullhorn said, "I'm Blackfoot. And besides white folk did most of the scalping of my people. You really bagged a deer in the scuffle?"

"I bagged a deer."

"I'm buying you a beer."

"Aaaanyway," Rooney said, "that's not the most exciting part. Abe here got ahold of the mine's dynamite—"

"Pretty red little logs—"

"—and we blew their water plant sky high, water every which way. You'd have thought the world had just learned how to rain like in the days of Noah, when the water before had only risen up from underground. Them scabs hid behind railroad ties and rail cars and regular cars and tough timber like we'd hid before. Well we had them trapped, we did, trapped like squirrels in a no-kill trap with corn down in it."

"The kind made of tin," Abe said.

"Then one of the guards tied a white apron to a broom handle," Rooney said, "and he came out from behind all that timber and steel. 'I wanna talk to your leader,' he said.

"'What do you want?' your Grandad asked.

"'We'll surrender if we can come out unpestered!'

"Your Grandad said, 'Come on out and we'll help get you out of this here county.' He meant it too, but he didn't mean it like they took it. There's more than one way to move a body. Well they threw down all their arms and came out and formed a line, and hundreds of us came forward with rifles and pistols and a couple of camping hatchets and we lined them up two-by-two just like with Noah's ark. One of them ran back to get his revolver, but I took his six-shooter from him and said, 'You won't need that where you're going.'

"He said, 'I thought purgatory's a frontier town.'

"I said, 'Get.'

"We marched up along the railroad to Herrin and after a good long while we had them doff their hats and put down their hands (they'd had them raised the whole damned time). Some of the union boys fired guns in the air. There was a negro there, armed with a hunting rifle running up and down the line and frothing at the mouth like rabid old Achilles, he was so damn angry, and we told him, 'Use your fists, John! Use your fists! See there? These here white sons-of-bitches? We don't think as much of them as we think of you, colored boy!'

"Your Grandad said, 'The only way to free Southern Illinois of all these nasty old strike breakers is to kill them all off and stop the breed.' I think that's the phrase he used, *stop the breed.*

"And I said, 'Don't you rush on into things, Patrick. Fools rush into things. Don't go too fast. We have them out of the mine, now, let them go like you said.'

"Grandad said, 'I'll turn loose when my jaw tells me to. Hell you don't know nothing, Rooney. How long you been working here? A day? I've lost sleep four or five nights watching them scab sons-of-bitches, and I'm going to see them taken care of. I'll show you how to work them big companies, you'll see.'

"Well, things got ugly, hitting them, the dust caked on their faces and cracked like the roads and dirt you see out west when it never will rain, caked and cracked like pie crust only it's filth and not sweet bread, and we came to Moake Crossing. Old McDowell bled from his head wound. Blood like a spicket dripping. Had him a peg leg made of old cork wood — so it was light, but sturdy — that made it really hard to keep up with the rest of us, falling behind the line like a broke caboose. Well someone offered to hang him to tie up that loose end, but we kept jabbing him with rifle barrels instead, not really wanting to hang nobody, and he'd get up. But then finally at Moake Crossing he couldn't take no more, stump was sore like it'll get on a peg leg, and so your Grandad said, 'You bastard. I'm gonna kill you and use you for bait for them other scabs.' He took him and another man down a crossroad and he shot him."

"Wait, wait, wait," Remmy said. "My Granddad shot someone?"

"Shot him right dead," Rooney said. "It was already getting that bad. Then someone shouted, 'There he goes, your God-damned superintended. That's what we're gonna do to all of you fellas too, that's how we're gonna do you.' That's when it got real, real bad. We killed so many of them out in the woods, killed them like little boys melt ants and army men, helpless little toy things running away, like how we shot up into them evergreens full of starlings every year and they'd fall down in clumps. There was this stout old fence with strands of barbed wire down there. Long one. Cut right through town back when they started enclosure and breaking up common land into parcels. Your Grandad said, 'Here's where you run the gauntlet. Now, damn you, let's see how fast you run twixt here and your homeland Chicago, you damned gutter-bums!' Then he fired."

Remmy didn't say anything. *That's* why Granddad Patrick never talked about this. Remmy'd wanted to learn more about his granddad, but not like this...

"Then everybody fired into them," Rooney said. "The mine hired guns dropped and the foremen dropped and the scab miners dropped and they put bullets in people still breathing and bullets into people not breathing no more and others tried to run *through* that barbed wire fence, and then their clothing and flesh caught in it like some sick leavings after some dark fashion show and if they was caught in it like a web and still living? Well in that sort of case we shot them twice. Someone cried, 'By God! Some of 'em are breathing. They're hell to kill, ain't they?' Out from the barbed wire some of them scabs later came and we went and hunted them down like dogs onto rabbits, hounds and foxes, hounds and coons through the woods all night, shooting and hanging and clubbing and hewing with them camp hatchets. One of us pulled a pocket knife and slashed the throats of those still living. I think that was a teenage girl or something, I don't quite remember: the memory itself is tinted red in my mind. Some men pissed on the bodies and then some ladies squatted and joined them. One man was dying and a young woman was standing over him holding a baby said, 'I'll see you in hell fore you get any water.' She put her feet on him and the blood bubbled out from his gunholes kinda like an old row boat, only it was death and hell he was sinking down into. Middle of summer's hot in Southern Illinois, you know. Well before the dark came on, them dead bodies had writhing black blankets of flies on them like you might see with ants and red velvet cake."

"Oooh… God," Remmy said.

"But that wasn't the end of it. Chicago turned against us since there was one of their big britches newspaper men down there and the *Chicago Tribune* got all the oil and coal money anyways — paid per ad, you know — so they ain't never gonna say too harsh and therefore too true of a word about them oil and coal companies and none of us have the money to buy an ad and tell our side of the story, so screw them. But them papers started shaming us for what we did. The papers in St. Louis turned against us, and once they did, the stampede started. We had Los Angeles papers and New York City papers and Texas papers — of course the Texas papers, what with Texarco — all talking about what godawful hellions we all were. After that, most of our senators and politicians turned against us, which made sense seeing as how they're on the payroll of the coal lobby at the time: they just needed a good excuse. Then the Governor and the President of the United States turned against us and tried to shame us. They didn't care about us, you see. They didn't talk about how profit's the only thing companies like that care about. How they don't think ahead for the life and health of their workers, the high water, the rooms with poison air, the buck fifty a day before the unions. They didn't want unions.

"I'll tell you, though, ain't a one of us down here's a red-blooded communist. Ain't a one of us down here that don't love America as much or more than any Yank New Yorker or richass Democrat banker or richass Republican real estate

Mogul in Chicago. We just needed someone to stick up for us, you see. And as corrupt as many of them unions got, ours never did. Them big city folk didn't care none. They wanted us all to hang. That's what we wanted too: we wanted all of them scabs and big money coal men to hang for shooting *our* boys and stealing the land *we* lived off of. We were just starting to come to the Native American way of things, you see, sharing with one another and thinking of the land as something we have in common. But it didn't matter then, cause every paper in America turned against Williamson County, Southern Illinois, the glory of Little Egypt and their unions. That was the last day of the real Little Egypt. We'd won the battle but lost the war because of their war on us with the papers. We need a paper that talks about us, truth be told.

"They had a trial for us. They burned through two-hundred-and-twenty souls just to get a jury that was favorable or likely to be — wasn't no one in Southern Illinois that didn't see the truth of it. Every member of that jury was from a big city or out of state or hired vets that protected that coal money and that oil money. Well they found the jury and they had a trial and that hand-picked jury found all eight men who had been accused of murder not guilty on account of the killing of Grandad's cousin and the other two as well as the strike breaking. So we won that battle, but we lost the war cause the whole damn country called it an injustice and tied it to the unions, and bit by bit the unions lost and went away and the rest is history all over these United States. That's why Southern Illinois failed to be the real Little Egypt. That's why none of the riches that came out of this land went back to the people that live here.

And that's why we loved your Grandad. And that's why he and I moved up to Boulder in the Kaskaskia river valley all them years ago to settle down and start our families."

"Bloody Williamson," Abe shouted up at the ceiling from the floor.

"Bloody Williamson," Bullhorn whispered into his beer.

"Bloody Williamson," Remmy said with brimstone and heat in his eyes.

They looked at the boy.

He looked at them and grinned that devil's grin he'd long ago gotten from his bloodline.

WILSON REMUS

1954

WISHING HE COULD buy one more day of laughs with Grandad Patrick, Remmy started to save in earnest. And he also started running around Centralia because Boulder and Carlyle reminded him too much of Grandad. The dating wasn't going so hot because she would only go with him sometimes, playing hard to get even long after she'd said *yes*.

"She" had a regal name: Elizabeth. Elizabeth Donder.

You bet your sweet patooty.

I mean she's the *very* one who'd been a little older than him and never winked back at him no matter how much energy he'd put into winking her way, back in those one room school days where the kids from the county meet out in one place and went to the same school that counted turnout in kids and oil folk, back when he'd learned about the war. Elizabeth Donder. Beth.

Elizabeth was his queen.

...or would be.

(He hoped.)

One of them times he tried to get her to go on a date with them even after they'd been going steady three months. "Will you go on a date with me? To the racehorse tracks?"

"Depends," she said.

"On what?"

"The news."

"The news? Hell, Beth, what I gotta do with the news?"

"There a curse in the news?"

"No ma'am."

"Then why you cursing?"

Remmy thought about that. "Well what kind of news will it take?"

"Give me whatever comes to mind, sonny."

"Well, we beat up old Jim Johnstone the other day, me and Pete."

"What the hell'd you do that for, Remmy?"

"Is it a curse?" he asked her back.

"Beating somebody up for no good reason is!"

"I can't win with you, Beth."

"Well why'd you beat him up, now?"

"Cause he was making fun of how poor our folk was and how rich his daddy was, and he threw a rock at me so I went over and whooped his little ass and said, 'The poorest carpenter's still richer than the richest oil folk.'"

Beth shook her head, her mouth wide.

"What?"

"You forget yourself."

"I forget nothing about myself. I'm five-foot-five and meaner'n hell: I'll grow more inches with my musky smell."

"No. You forget my daddy works for Texarco and *I* am oil folk."

"Well... well then, I guess I date oil people, and that's alright with me."

"Not tonight you don't."

"What?" he asked. "Why the hell not?"

"Cause you're mean and you stink and you should go take a five hour bath in a galvanized tub."

"Well that's mean, telling a feller he stinks."

"I can smell the bullshit coming off you a mile away, my man."

"We're not done dating?"

"No sir."

"So will you go with me to the races?"

"No sir."

"Well, shit. How bout if I bathe?"

She smiled. "We'll see. You'd better straighten up, Remmy. You'd better."

"I'd better. Sure. Anything you want."

Remmy knew Robin Hood needed a Lady of the Woods to keep him straight and narrow and to compete with him on adventures. But Robin Hood also needed his dangers and impune pranks so he saved up in a jar for them and man that jar got threatened with fines. One night Remmy rolled up to a stop sign and this other guy had a car just like him. Remmy had the older car.

This guy said out his window, "Let's race."

Right on Broadway. Four blocks away from the railroad tracks.

They were both doing about 80 miles an hour when they hit the raised earth of the railroad and took to the air. Both cars smashed down in sparks: promises of fire.

And here came the police, them ocean-blue lights.

Remmy ran a few more blocks and decided to stop. He had the car so he got out and faced the cop like you ain't supposed to do these days. Sheriff Chubb Sanders, later to be Police Captain, looked at Remmy's I.D. and the wheels started turning as he said, "Broganer… Broganer… Oh hell you're John David's son. What in the hell are you doing out here racing like that and you're going to get someone!" It had started out as a nice question anyways.

Sherriff let it go like it had been Remmy's dad's car. But Old Chubb talked for an hour. On and on and on, chewing on him. "Who is that other fella?"

"I don't know," Remmy said, truthfully enough. "I know he's from O'Fallon or west. Comes over here all the time. I don't know his name."

"You know him."

"I honestly don't," Remmy said.

"You know him, and a good man would turn him in if he knew what was good for him."

"I don't know him, Chu—"

Chubb Sanders eyed him.

"Officer."

"You'd better turn him in."

"I will and gladly, soon as I find out who it is."

Chubb Sanders watched him and finally said, "Get back to Odin and stay out of trouble."

Saved Remmy a big fine, his father did. Would have

cleaned out his prank fund. Hell he wore out the knees on his pants begging like that.

That year he had a '48 Chevy, and the bumper came from Sears and Roebuck. Had pushbuttons on the dash. You could turn the dial or punch it for six stations. But somehow if you'd turn the dial and punch the button at the same time, it would jam, and Remmy got the police station's dispatcher. So he knew where the police was at. It came in handy, that. He could run a stop sign or drive too fast sometimes. Had a lead foot once in awhile, and it helped cover for him. At the Central City skating rink, the one with the white laced roller skates and the big old painting of a beaver, he knew where they were at and could get to a race strip or to pulling a prank outside of their view.

He worried Beth might find out. God help him if she ever found out: he'd fare better in prison.

Thing about the Magic Chevy was it actually had the heart of a Buick. Remmy didn't find this out later, but the starter was Chevy and the engine was a Buick. Good thing too cause Chevys still used that splasher system for the oil — little scoop on the bottom of the rod bearings that dragged through a channel full of oil and forced the oil to the bearing. You go to that engine too fast, that channel didn't fill up, bearings went dry and then *melted.* Chevys didn't have no bottom until the V-8. So he turned out pretty damn lucky that someone kept the skin of the Chevy and the heart of a Buick, cause he raced the shit out of that thing and would have melted her soul. That's the other reason she's magic.

One night they went racing again, and the police went after Remmy and his trusty steed, and he decided to use the car's magic power. He dialed it and pushed it. They came after him, but he knew they were going down east McCord in Centralia near those tracks again, so Remmy turned right before the police came over the tracks. He gassed it down that alley, pulled in the empty garage, shut the lights off, and waited.

They passed twice.

Then Remmy got him and his friends and his Magic Chevy outta Centralia for good.

"For good?" Pete Taylor asked.

"Cars'll get you in trouble almost as bad as girls."

"Well hell, Remmy," Pete Taylor said, "I think hanging out with me'll get you into far more trouble than cars or girls."

"What are you talking at? *I'm* the one that'll get *you* in trouble."

So they argued about who's ornerier.

Truth was, the cars might've been why the girls played so hard to get with Remmy. All of them except for Elizabeth, that is. A man's only made for so much chasing.

And Beth still hadn't found out about him and his racing…

"Come to church with me," Beth said.

"I'll go to church with you," Remmy said.

"And Sunday School," Beth said. She was working at the bank by then, see. A banker. Not a movie ticket seller no more, a bonafide banker.

"I still go to Sunday School in Odin," Remmy said. "I

still remember the songs they taught us as kids. *Only a boy named David. Only a little sling. Only a boy named David but he could pray and sing. O—*"

"Fine, just church?"

"I said I'll go to church with you," Remmy said.

"Alright."

He went to church with her and tried to hold her hand, but she wouldn't have it in church.

People giggled at them from the pew behind them. Little people. Kids.

Remmy felt uneasy. He was used to a different church and everything was all off. The order, the hymns, the prayers, the other stuff. It wasn't bad, it was good, it just wasn't the flavor of good he's used to tasting. It was an oil-folks church, too fancy by half.

This One Lady came dressed up really fancy. Too fancy. So fancy you could tell that church was the only thing This One Lady had left to dress up for, not that you had to dress up for anything, but dressing up for a dance can be fun in its own way. This Really Fancy Dressed Up One Lady walks up and looks down on Remmy. "Boy that's my seat, get out of there!" She was loud and mean. In time, she'd become Dr. Gabriel's mother-in-law. It's a wonder she didn't use some other words. If Remmy was gonna save up in his jar of prank money for Sherwood Forrest, for buying up paradise by buying a church someday, it wasn't gonna be that one for damned sure.

He asked The Good Lord if that was his idea of a good church.

Good Lord said, "No, Remmy, I'm not really proud of that bunch, but I love them something fierce."

"Me neither," Remmy said.

"But I love them something fierce."

"Me neither," Remmy said, "but I think I'm about to be related to some of them."

"You're not listening," The Good Lord said.

"You're not listening: I'm about to be related to one."

"Try being related to all of them," The Good Lord said.

"Good point."

Remmy never sat in This Really Fancy Dressed Up One Lady Related To The Good Lord's seat again cause you got to be careful. She was pretty wild. Her blouse colors alone...

Well that winter it was so, so cold that there's snow on the ground and ice and it took forever to get the car started. He grabbed a washer and an old crackerjack diamond and glued one to the other and went driving with Beth and faked like his Magic Chevy broke down on the road cause of a flat tire on Beth's side, so he got out of the car and walked around to the tire and started hitting it with a tiny wrench he'd brought along. This little *tink* and then *tink-tink* and a pause and a *tink*. He knew the thing would annoy her eventually.

Beth rolled her window down. "Are you okay?"

"Could ya get out and help me?" he said. "I need an extra set of hands to hold something."

"I'm in my nice dress."

"Well it's not an oil can. I just need you to hold something real quick like."

"I might get the hem dirty," Beth said.

"Then you'll be on an adventure with me. Come on!"

"No! We got to get through this snow. Fix the tire."

He groaned. He wasn't ever gonna get to propose to this

woman, how stubborn she was. He slammed the wrench into the wheel and shouted, "I FIXED IT!" even though there's nothing needing fixing.

"Hurry," she said. "I gotta use the lady's room."

He chuckled to himself.

They drove for awhile through the ice and snow and cold.

"Hurry," she said. "Hurry please."

"Well you don't want me to wreck, now do you?"

They went for twice as long as the last stretch.

"I'm not gonna make it, Remmy, pull over."

"Yes ma'am. What you gonna go in the woods?"

"Sure," she said. "Keep this car running and warm."

"Yes ma'am."

She hopped out and walked towards the trunk.

He turned on the radio and waited. Snow fell in whisps like Momma Midge's old snow globe, whipping round as fast as he could crank it.

Beth didn't show.

He changed the station. Another six songs played.

Still no Beth.

He leaned over to her side of the car and rolled down the window. "Beth?" he called. "BETH?!" He shouted out into the woods.

A small voice way closer than he thought said, "I'm here Remmy, no need to go on shouting."

He looked over to his blind spot.

She was leaning up squatting against the car.

"You okay?" he asked.

"No," she said.

His heart went to hammering. "What's wrong? What can I do?" He was worried.

"Well… I…" he heard her gulp louder than a damn cartoon. "I'm stuck."

"What like your knees got weak?"

"No."

"What do you mean you're stuck?"

"I… well when I squatted down to make water… I leaned up against the car."

"You didn't. Buckassed like that?"

"Yeah."

"HOLY SHIT YOUR ASS IS FROZEN TO THE CAR?"

"Yeah. Don't shout, please Remmy, oh God I'm mortified."

"Well hell you're gonna freeze, Beth!"

"Don't I know it."

"Don't freeze for shame," he said.

"Well what are we gonna do?"

"I got an idea."

"Oh God, don't you get an idea. I'll think of something first."

"I got an idea."

"Just give me a minute."

He gave her a minute. An old jitterbug song came on. It ended. Another began.

"I don't have any ideas," she said. "And I'm getting cold. What's your idea, Remmy?"

"You're not gonna like it."

"Just tell me, I'm freezing."

"I'll pee on you."

"You'll what?"

"I figure you need hot water to melt the cold and if you pull hard right as I pee on your ass, you'll either get off or be frozen for good."

"Oh God no."

"You got any other ideas?" he asked her. They ain't even kissed yet.

He waited.

And waited.

Waited until a Johnny Cash song came on. Ring of Fire. He giggled thinking about the lyrics. Twice.

"No. No ideas."

"Okay I'm coming," he said.

"Wait," she said.

"Okay," he said.

"Promise you won't look at my bum, but just at the car metal above it."

"Long as you turn your head and don't look at me while I'm going."

"Oh Good God Almighty why did I ever start dating you."

"Cause I get you unstuck," he said as he grunted and exited the car.

He walked right up to her and forced himself — very, very hard — not to laugh. She looked like someone caught shitting in the seat of some carnival ride. "Don't you look now."

"You neither," he said.

He looked in her eyes. "Okay, turn your head. You're gonna need to lean as hard as you can towards the snowditch."

"Okay," she said. She pulled and turned away.

He looked straight at the place above where her butt touched the sheetmetal of the car, unzipped himself, and started peeing. Shouted, "Pull hard!"

Cracks sounded as the ice gave and she fell face first in the snow. He looked at her and turned his head, ashamed. She shot up, pulling up her drawers, and let down her dress, staring at him. "Remmy…"

He looked down and pulled up his pants. "You said you wouldn't look!"

"Oh don't tell me you didn't sneak a peak when I fell."

"That's beside the point."

He walked right to her, got down on one knee in the yellow snow, pulled out that glued ring and said, "Will you marry me?"

She punched him. Hard. Twice. Three times. She kicked him in the knee.

"OW! God, lady, I just saved you from certain death!"

"I coulda saved myself!" She kicked him again while he was down.

"Fine!" he said and started to get up.

Then she said, "Yes."

"Yes, what? Yes you could have?"

"Yes I'll marry you."

"Well Good Lord, I thought that was a no. Why'd you punch and kick me?"

"Cause my birthday's in two weeks, and you're supposed to do it then not while peeing on me to get me de-iced."

"Now you get two gifts," Remmy said as they got back in the car.

"Well that'll be nice," she said.

Shit. That meant he'd need to buy her another one.

"I hope this isn't my real ring," she said.

Two other ones. It was no use arguing with a banker. He'd have to dip into the fund. But for his Maid Marian, it was worth it cause every once in a while, a sale was so good, it actually did make you more money in the long run. The bankers called those investments.

And she was the greatest and best investment a man could ever hope to find.

WILSON REMUS

1954-1955

REMMY SET THEIR wedding day for late December to save some money towards his master plan. They borrowed trim from the oil-folks church and a dress and a suit coat and got their groomsmen and bridesmaids to really come together and make it easy on their plans. Ain't nothing better than a wedding stitched bolt by bolt of tulle by a community that loves you more than any naysayers who don't want you and your wife getting together, and that's The Good Lord's honest truth.

And then a week before the wedding, Beth brought it all up. "Well, you started dating me after Christmas, and you proposed before my birthday and so you've never given me a gift at all!"

"That was my plan," Remmy said.

"That ain't right."

"Well, maybe not." Remmy thought of Grandad Patrick and his generosity. "What would you like?"

"I'd like a dozen white roses," she said.

"Oh boy."

"And a dozen red ones too."

"Oh boy."

"And I'd like you to buy a nice thick ring for yourself."

"That's not—"

"A ring for yourself so that everyone knows you're my man and nobody else's."

"Oh boy."

"What's THAT *oh boy* for?" she asked.

"Because if that's why you want it, there's no way I'm not getting it."

"Good answer."

"So oh boy. That's a lot of money: a ring and two dozen roses white and red." It was okay to dip into the fund for this, wasn't he twenty years old now? He figured kings had roses at their weddings and wore rings. Rings with their signatures on it. That's what he wanted: a John Hancock ring, only one that had a big old R on it.

Now about this wedding. There's only two things you need to know. First, the Preacher had been in the navy. Good Preacher. Lotta people got married at Young's Chapel cause the preaching was so fine. This Preacher loved the navy but he played cards all the time in the navy like you do. He gambled away most of his earnings and he struggled for most of his life, only getting food when he was starving and hosting as many potlucks as he could. You know the kind.

So Remmy was back in the back with him, waiting around, and the Preacher asked, "You shoot dice?"

"I do when I'm this bored," Remmy said.

"You're about to get married," the Preacher said.

"I like the married part," Remmy said. "It's the *about to* that has me bored out of my gourd."

"Oh. Well you want to shoot some dice?"

"Okay." Remmy didn't never gamble on account of the fund and his ability to do math and figure his odds. But he also hadn't never gotten hitched, so he didn't quite know how to pass the time good and proper like. Gambling was as good a way as any kind in his mind.

The Preacher said, "We'll bet a dollar."

That seemed like a lot, but Remmy said, "Okay."

They rolled. Remmy made his point and won.

"Double or nothing," The Preacher said.

"Okay."

They rolled. Remmy won. He was up to two dollars.

"Double or nothing," The Preacher said.

Remmy thought about his fund and said, "Okay."

They rolled. Remmy won. He was up to four.

The Preacher swore.

Remmy chuckled.

"Double or nothing," the Preacher said.

Remmy thought some more and ran some figures. It still worked out, so he said, "Okay."

They rolled. Remmy won. He was up to eight dollars which was getting close.

"I tell you what," Remmy said. "I'll give you one more chance to win it back and then some. I'll bet four dollars more, which is twelve dollars."

The Preacher said, "Okay."

They rolled.

A die fell off the spare altar, the one they were rolling

on top of. The altar hid in the back except for big events like passion plays.

The Preacher caught it and rolled it again, cause ain't no die roll counts that goes off the table, specially when you're casting lots on the altar. That's a universal rule.

The die landed and Remmy won.

The Preacher swore.

"It's okay," Remmy said. "Twelve dollars was what you was gonna charge me for the wedding, and I think we'll just call it even there."

The Preacher was red-faced, but they shook on it and the preacher said, "Consider it a wedding present."

"Oh I do," Remmy said. "And I thank you for it." He looked over in the corner by the spare altar at the crumpled up pile of Jesus clothes and Bible costumes from the passion play they'd done and said, "Huh."

After that those two and the others had to lift a couple more things changed around for the service, and one of them was this old bit of oil pump metal in the back. It slipped while Remmy and Pete and the boys were lifting it and it cut his hand really good. There was blood all over the back of the church.

John David had brought the first aid kit. He wrapped his son's palm in the cloth and put the last of the merthiolate on it, and it burned.

"Hah," Remmy said, even though it was burning.

"What?" John David asked.

"The last of the merthiolate. You told me we'd never get through all three tubes."

John David raised his hands, the empty tube in one,

and said, "How was I supposed to know I was raising two hooligans?"

"Best penny I ever spent," Remmy said, and nodded to himself. "Buy one tube of merthiolate and get the other for a penny."

"What other family in history ever used it all?" John David shook his head and went back to winding his son's palm.

Remmy thought about a book he'd just read. Old Hemingway'd written *Across the River and Into the Trees,* book about facing your own death. He stared at his hand. He stared at it and grinned a devilish grin.

Beth ducked in. "Will you fools get to your places? I'm the one that's supposed to walk down the aisle."

"We could change the song just for me," Remmy said.

"What song?" Beth said.

Remmy sang, "*Here comes his hide, all tanned and fried. All cause his bride spanked his hide when he lied.*"

"What did you lie about?" she asked.

He could see he'd just made a mistake. "Nothing."

"Remmy!"

"Gambling."

"You did what?"

"I won us a free preacher rolling dice," Remmy said.

"Well..." Beth said. "Well... I guess the Good Lord works in queer ways. Don't tell nobody this happened."

So, of course, Remmy told everybody. Then he said, "Good Lord, you have anything to do with this?"

And The Good Lord said, "Obviously your soon-to-be-bride hasn't read Proverbs sixteen thirty-three."

"What would that be, Good Lord?"

"Verse about shooting dice," The Good Lord said.

Remmy grabbed a Bible and looked it up. "Well I'll be damned."

"Not on my watch," The Good Lord said.

Remmy asked The Good Lord, "You got one in there about the defrosting properties of hot urine?"

"Remmy!" Beth said.

"I'm coming."

"He just never listens to me," Beth said to Gwen.

Gwen said, "Mine listens but never talks back. Maybe we should swap an ear apiece and half a tongue."

"I got some real good shears," Beth said. "Course if we did that, you'd be kissing your brother half the time."

Gwen shrugged. "I'd trade half incest just to get mine to talk."

Beth raised her eyes. "Really?"

"Good Lord no. I don't know how you kiss that brother of mine."

"Cause I smell like the salt of the earth and look like the light of the world," Remmy said.

"Well hell," Gwen said, "you're both cousins."

"Not really, sister," Remmy said, "it only counts if you marry your aunt's daughter. Other than the last name Donder, we ain't got no idea how we're related. And besides, it's like I said: I smell like the salt of the earth and look like the light of the world. Ain't that right, Reverend?"

"Whatever you say, Remmy. I won't cross your good fortune again today."

The girls all had purple dresses cause there'd been a sale on purple dresses, and they couldn't have dresses otherwise. Problem was, they were all sized same, so you had to get the girls in there one way or another, squeezed or stuffed. One of them was this really big girl who walked on tiptoe like a ballerina, and it worked okay until the hoedown when she bent over and it turned into a stripper dress. Even the Reverend got to blushing then, dice or no.

But back at the ceremony, it's only a handful of people pulling together with whatever they had in that nine-pew chapel. Preacher preached about, "My rest I leave you, my rest I give to you," and about the bride of Christ and the Bridegroom.

Remmy leaned over. "You realize, that makes me Jesus in the affair."

After a moment, Beth leaned over. "That means you got to die for me."

And Remmy leaned over. "Yes, but that means you got to submit to the victory of my death."

And Beth leaned over. "Yes, but it also means you're gonna be a corpse three days for it."

They went back and forth like that, talking about who was gonna submit to whom and who was gonna die for whom until the whispers turned to loud talking and then shouting and then an awkward hall of folk watching Remmy and Beth carrying on.

They stopped cause they heard a soft sound before them.

It was the Preacher whispering, "Can I finish my sermon?"

And they both looked at the blushing crowd in the pews.

And the Preacher read Ephesians instead. "You both was escalating the wrong way. Remmy, you gonna die for her?"

Remmy nodded. "I'd do anything for this girl and she already knows that's the truth."

"Beth you gonna submit to him?"

She nodded. "I'm gonna serve him so hard and respect him so much, he's gonna want to fire me."

"Good. I want a nice clean marriage, no shots below the belt."

Remmy said, "Obviously you've never been on a honeymoon or stuck on the side of the road."

Preacher's turn to blush again.

Beth did struggle to get the ring on his finger on account of the gauze. But she got it done in the end.

WILSON REMUS

1956

THEIR FIRST HOUSE was no castle — wasn't even a nice fort out in the woods somewhere from which the merry men could stage their prank on the Sherriff of Nottingham – but they couldn't afford much more. Besides, he needed to find more merry men before he could do all of that. It was a four-hundred square foot, maybe less. Fifteen dollars a month. They had an oval number three tub, and you'd heat the water on the stove if you wanted to take a bath. Had to get it just right, between boiling and the well-cold, but you could after some practice, you could.

Beth learned to cook in that little shack. Cobbler and meatloaf and other things that Remmy liked or grew to like or didn't grow to like but pretended like he had. Mostly. Sometimes he could get mean about it, specially at first, but he got better. That's what marriage does, you know, holds a mirror up to every crack and crevice it finds hiding stuff down in your soul. So he got better over time at being better to her.

You didn't have a toilet in shacks like that back then. Sure the fancy schmancy apartment towers in New York City had toilets, but not the folks out west of the Great Lakes and south of Highway 40. Well except the ones with the ads in Times Square — he bet they only sold the outside for ads and the inside sat nothing but squatters.

A privy. That's what you had. An outhouse. Not for just camping at the state parks. For your four-hundred-square-foot shed like Remmy had. Oh don't act like it was torture or anything, this wasn't the back of some Greyhound bus used by fifty thousand people before the guys clean it. Remmy and Beth kept it clean and scrubbed from bugs and spiders and dirt, kept a little pail of clay handy to toss down the hole after each using to keep them smells down. Sounds like nonsense, sweeping a dirt floor clean, but you can do it if you know what you're doing. After all, a granite countertop's just polished dirt. So's all them statues in the Metropolitan Museum of Art. It's the work put into the carving and the polishing, not the starting dirt, that matters.

In fact, the best dirt is really just a big old pile of shit and that big old pile of shit hardens over time by leaking out all the soft stuff, and the hard stuff became gems and minerals and things, and the soft stuff became oil. One you could polish for making your home and city pretty and the other you could dig up to power your cars and boats and planes. The whole fancy big city world was just a well-polished, well-designed pile of shit that ran on more shit. Bullshit powered bullshit. You simply had to think about all that shit in the right way to make it look pretty and work good. It's the mind that matters, not the body or the will.

Wasn't it God himself that put his breath into dirt to make a man? Dust we are and to dust we do return without His mind giving us leave to be in the between time. It's the work put into the carving and the polishing, not the starting dirt, that matters. So yes, they minded that dirt floor and therefore kept that dirt floor of that privy pretty clean of dirt and spiders and bugs.

Problem was, they lived right next to a service station out there by the oil fields. It was one thing for that sour sulfur smell to stick on everything, what with the flares burning off God knows what from the tops of those pumps. But the service station right next door made it hard to have a privy. You see, people got to using it at night when they stopped for gas — service stations didn't have no plumbing then, either. Remmy parked the magic Chevy between the service station and their house, the only barrier making safe their property. But customers'd walk around the car and use that privy over and over again. They'd track in their muddy boots and track out their smells again, or leave the door open which meant there'd be coons in there picking through the trash. People used Remmy's privy so often he wondered if this was the kind of throne that kings and brigands had to deal with: half the kingdom coming in to use the garderobe without asking.

Remmy got sick and tired of cleaning up after all of those strangers, so he took a nice hinged latch and threw a padlock on there, a nice combo lock so he could share the combo with his neighbors and some other folk. He wasn't an asshole, Remmy, he just didn't want every living soul in every county south of Lake Michigan using his privy.

Well one night he had a pain. It was enormous. He

didn't want to wake up, but he knew there wasn't no running from it. It was cold, so he put on his long underwear and his pants and his other shirt and his coat and his hat and his scarf and his gloves and damn it was taking forever. He walked out the door and went straight to his privy, the shine of the service station lights warding him as he waddled and winced, how it hurt him.

Up to the door of the privy, he called out the numbers as he turned the dial. "Thirteen left. Four right. Twenty-five… no, no." He spun the dial to start again.

"Thirteen right. Dammit. I need to make this a key lock."

He had to go so bad, he crapped his jeans and long underwear. That's why they make them long, you know, just in case. Standing there in his own shit, having locked out strangers and having so locked out his very self, there in the light of the service station, he reacted the only way it made sense to react. He got the giggles. Sure it was funny. But the funniest thing to him wasn't the obvious thing. He thought it hilarious that, as bad as he stank in the midnight winter, he still couldn't cover up the stench of the oil fields burning those flares: sulfur bombs like Beth's farts under the sheets.

King Arthur sure as hell never had to worry about this shit.

WILSON REMUS

1957

THAT OLD SHED tired them out, living in squalor like that. Or feeling like they lived in squalor because locking travelers out of your privy ain't what you do if you're living in luxury. You either welcome the whole town or lock them out of the whole property, but you sure as hell don't padlock nothing but your shitter.

Beth talked and talked at him, and finally Remmy agreed: they went to move into a place that he could fix up. There were a bunch of Frisco houses there owned by Texarco, and carpenters like Remmy would move in and buy them up. Fix them up. Good friends, too, some of them carpenters. And others became good friends in time because they were good neighbors. It was almost like having his own merry men, living out there like that. Old Sinclair, who played in the stone and mortar and concrete. Ryan who'd later get pretty good at shooting things with a nail gun, even though the nails turned end over end. Pete Taylor, big as Little John or Sir Bors. He had friends with last names like Cooper and Fletcher and Tanner and Smith, Baker and Taylor and Gaffer and Knight — the kind of last names Southern Illinois'd forgotten the meaning to, but Remmy spotted as the descendants of ordinary people known by their work. Men made merry by the work and the folk it was for, how a cooper laughed for his winos, how a fletcher laughed for his hunters, how a tanner laughed when your father caught you lying and tanned your hide. Merry men. These were the ones he'd need for his great prank, his way of taming that tyrant as did every good wit and jester.

There on that block, they'd have potlucks and cookouts and sprinkler days for the kids (though Remmy and Beth didn't have any yet) and they'd have garage sales in their neck of the wood to make a little money with sales bins. If they'd have been in the army together, they might have scarred their skin with tattoos, but being carpenters and day laborers together, they didn't need no ink and no needle to have matching scars. Didn't matter: skin and mind ain't the same. Sure it wasn't Camelot yet, but it was sweet. Working for Texarco and living among them and coming home to Beth's cooking. In later years, they'd drive out there and show people where they got married and how they'd started out, and Beth didn't really like that, feeling shamed and all, but Remmy was proud of it then, and he was proud of it later on account of the joy they shared.

At the time Beth worked for another bank, you see. Nice guy, but it wasn't the banker Remmy'd worked with for twenty-two years (including the time under his mom and dad), so he was wary.

Remmy didn't like it much, but he borrowed $1,000

for that other house. He had to dump out the whole CAMELOT, MY MERRY MEN, AND OUR PRANK jar in order to put anything down at all. What they'd wanted was a nice house out in the oil field with mineral rights, but Texarco didn't want to sell those to him or to John David. The banker wanted John David to cosign, so Remmy said, "No thank you." And he went to Beth's boss.

Beth's boss? He set up a nice deal:

Remmy paid the bank $25 a month.

He paid the lumberyard $25 a month.

And they moved out to that house beside all of the other Texarco carpenters that weren't allowed to buy out in the oil field but were given all these shanties. The walls were falling off of the bathroom. It was bad.

Yeah, they showed them in later years where they started and people was unimpressed. Instead of having a commercial, septic system Remmy made their own mainly using two-inch pipes.

"You need a four-inch pipe!" they'd shouted at them.

And Remmy found out why when the shit backed up now and again. But two-inch pipe was all he had and all he could afford. Even still, they bought that house for half-price just like most of the carpenters in the neighborhood had done. Out front of the house, there's this great big ditch. A six-foot wide ditch. And over that ditch stood a little bridge built out of wood.

Few weeks after Remmy bought the house and started repairs, the Texarco folk came out with a big old tow truck and pulled up next to that bridge.

"Hey there fellas," Remmy said from his porch and walked up to them.

"Morning Remmy, we won't be long." They started dropping the chain.

"The hell?"

"You bought the house, Remmy, but you didn't buy the bridge. That's Texarco's bridge."

"I beg your pardon," Remmy said, "if you put a culvert in town, it belongs to the city.

"Out here it belongs to the township. That means Texarco keeps it." The man turned to the driver and shouted, "Throw a chain on it!"

And the boy with the wrecker came and threw a chain on it.

"Now come on this is just silly," Remmy said. "What are you gonna do with my bridge?"

The man turned around, his whole face following his shit-eating grin. "Give me twenty dollars or I'm towing the bridge."

"This is the dumbest thing ever," Remmy said, "Not just cause of the situation but how you're going to tow a bridge? It'll throw splinters all over town."

"I'll figure something out, so you'd better give me twenty dollars."

"Dumb, dumb, dumb." But Remmy paid and got it in writing that the bridge now belonged to him and would connect his home to the road and the road would connect him to the wide world beyond where he could travel freely and come back to the door where he began, like a living water cycle. They left him alone after that.

At least about the ditch.

Later that summer, Daddy John David and Momma Midge took Remmy and Beth and Gwen and her husband

Ryan out to the Kaskaskia river valley, not too far from Boulder where Grandad Patrick had once chained up a young boy as a mail-order bride, not too far from where John David grew up. Not far because Momma Midge didn't travel outside Little Egypt. It was to be a picnic. Rita, Beth's sister, never came to these things. She'd moved to California or Oklahoma or something.

They had a really nice meal. They had fresh fish and John David had brought along a big old charcoal grill and had grilled these steaks and made little aluminum foil packets of potatoes and onions and carrots for the coals of the bonfire out there in the field. They ate good, they did, and told some stories there in the valley.

"What's the occasion?" Remmy asked his father.

"Look around. All of you."

Beth set down her romance book and looked out and the golden hour light made her glow like a bronzed statue of Minerva, bearing the tiniest firstborn child that Remmy had conceived in her only a month prior, a bronzed Minerva with the primordial child weaving together in the secret place. But Remmy finally tore his eyes off of her and looked around.

Sure was pretty. Pictures from that day, old as they are, look pretty even now. The copses of trees that grew up around the barbed wire, the country houses here and there, the churchyard off in the distance with the graveyard, the way the hills rose up just right, unlike most places in Southern Illinois — a wondrous land in which you could see forever and think on it too, if it wasn't for the trees. Little Egypt. Promised Land.

"Wow," Remmy said.

"Drink it up, gang," John David said.

"Why's that Dad?" Gwen asked. She was darning a pair of Ryan's jeans.

Daddy John didn't answer at first.

Gwen set down her stitches. "Why's that Momma?"

"It ain't good." And then Momma Midge spat. She never spat. But she spat then.

John David sighed. "Because next year I hear Texarco went and bribed the Army Corps of Engineers to dam the river and flood this place, and seeing as how I grew up here and remember so many good things, and seeing as how I don't think it's natural damming no river like that, specially one named by our Indian ancestors in a state named after them, I thought it would be nice to have one last feast here in the valley and its light."

That was the first time Remmy'd seen his father cry since the incident with the milk wagon when they'd treated each other like Bloody Williamson. The way the green light of those fields reflected in his tears, the blue of the skies and the oceans worlds away that even those skies reflect. It almost looked like the shepherd of the woods and fields had cracked and the waters of the deep had started to leak their way out of him.

His father muttered something and looked up, the eyes of him catching light.

"What's that Daddy John?"

His father muttered again. Remmy caught an "e" sound.

"I… I'm sorry, pop, I just… I didn't hear you again for the mumbling."

"Oh God Bloody Williamson!" he cried out. "Bloody Williamson! Bloody Williamson! Bloody Williamson!"

He didn't stop crying for a long, long time.

IV. Adagio

…Nor is it strange these wanderers
Find in her lap their fitting place,
For every particle that's hers
Came at the first from outer space…

—Jack Lewis

WILSON REMUS

1958

"I LIKE THE OTHER one," she said.

"But it's so heavy," Remmy said.

"I know. I like a sturdy table," she said.

"There's kitchen tables and then there's butcher's blocks. I don't plan on slaughtering a hog in my dining room."

"I might," she said. "I'm the cook, don't forget."

Remmy sighed. "Do you think it's pretty?"

"I do."

"Do you think anyone else will?"

"What do I care for what anyone else thinks?"

"I'm not going to answer that question, you'll get me into trouble no matter which way I go."

"Wilson Remus Broganer, what's that supposed to mean?"

Wilson Remus Broganer said nothing to his pregnant wife.

"Well in any case," she said, "I like it and I think it matches everything else so this is what we're getting."

"All right now, I didn't mean to make you mad."

"I'm not mad. I'm decisive."

"And you won yourself a decisive victory. Okay now, let's get it to the bag checker."

"They don't make bags big enough for this, honey."

"You know what I mean." He hefted the massive slab of wood. "You know what I mean," he muttered as he grunted and hefted it over to one of the counters with one of them Sears and Roebuck signs and they bought it and, with much struggle, took it home.

"Five dollars a month," he kept mumbling as he drove. "Five dollars a month. I could buy a share of Texarco at the end of the year."

"Well then, let's take it back."

"You were listening to me?"

"You weren't whispering, really, now were you?"

"I meant to be."

"You didn't mean very well."

He harrumphed. "Well, I'm glad you have your table."

"Are you?"

"It's just so big."

"I think it's pretty."

And so they went until they got home. He set up the table, and it filled that room as does the king's table in a great hall. Except this was no castle. Not quite yet. But once he saw it in place, he was pleased enough with it.

"You're smiling," she said.

"It's a pretty table set up like that."

"That's why I made you buy it."

"I just did a good job setting it up."

She did not return the affirmation.

Two weeks later, the biggest and baddest tornado any of them carpenters could remember hit. It snapped hundred-year-old trees in two. It leveled some houses and frayed some power lines.

When it hit, Remmy was at home with Beth.

"It's a semi-hurricane!" Beth screamed.

"Tornado."

"We're going to die!"

"We might," Remmy said.

"Why would you say such a thing?"

"I mean I'm sure we'll be fine honey. Let's go to the storm shelter."

Well they went outside and that wind blew everything. Tore at the trees. Tore at the roof he'd just fixed up. There were boards and water thrown every which way. Beth waddled a bit with her swollen body and the baby inside. They got to the storm shelter door that Remmy had just dug and they opened it up and damned if there wasn't two feet of water down there.

"That's not a storm shelter," Beth said.

"Sure it is, get inside!"

"That's a pool! That's an underground lake like in the movies where the monsters hide, I'm not going down in there!"

They were shouting over the wind, mind you. Stuff still blowing all around them.

"You get in there or I'll get you in there," he said. "We're not going back in that house! I know how it's made, remember!"

"You do whatever you want," she said, "but me and the baby are going back inside."

He watched her go, holding that sheet metal door like he was. Then he groaned and let it slam and he barely heard it over the cry of the trainrumbling in the sky. He started following her inside and then he shouted to no one in particular, "Where the hell did my brand new lawn chairs go?" They'd been metal chairs, heavy chairs, sturdy chairs, and not a one of them was in the backyard.

Back in the house, Beth was pacing, slowly, off-kilter. "What do we do?"

"You vetoed my plan!"

"What do we do? What do we do?"

The house had been built just on concrete blocks. Well the wind got under there like two stock boys will get under a box and it lifted the whole house up about six inches and slammed it back down again.

Beth screamed.

Remmy, for once, had nothing to say.

"What do we do?" She was begging, now.

The house was lifted higher and slammed down on those concrete blocks again. Some plaster fell off the ceiling in the other room.

The table.

Remmy went into action. "You get under there. It's new but it's the strongest thing we've got."

She got under there, moving like a station wagon with two missized wheels, like a wheel with a ten pound weight on one wall. Then she sat still and some plaster fell off the ceiling and onto that five-dollar a month Sears and Roebuck kitchen table.

Remmy didn't bother to duck under anything but one of the open doorframes, leaning against it like some cowboy watching his horse from the porch of a saloon. "Five dollars a month. Decent insurance policy, I guess, I don't know."

More plaster fell.

The house bounced once or twice more. Then the wind died down.

And then a sound like what you'd expect if Chicken Little'd been right and the sky really did fall. Something like the crashing of the Tower of Babel. Something like the fall of Troy or the breaching of Atlantis's levy.

Finally, all went calm.

"The hell was that?" Remmy said.

"Oh, Remmy, do you have to cuss?"

"I only cuss at cursed things and whatever just happened wasn't no blessing."

"Is that why you've been cursing at me?"

He didn't answer that. He walked outside instead.

There stretched out in his yard, discarded as if some god of greed had found no more fun in a playtoy, sat a massive tower. As if the angels had gotten bored with their scaffolding for building the pearly gates and had kicked it over the edge of heaven.

"Bethy come look! Come look, it's awful! It's an awesome thing to look at."

The neighbor carpenters and their families were outside too on their front porches, looking at the mangled black thing spread across the yards, the rain still coming down and none of them caring, not even the prettier ones in their nightgowns a bit too early—maybe his neighbor Joe'd been weathering the storm with a little bit of marital duty. How people do cling to one another in hard times, even if it ends up with your neighbor Joe's wife standing half naked on her porch.

Beth came out and looked with Remmy. They all looked at one another.

Then they all stared at the massive steel oil derrick, as long as a water tower is tall. His father'd told him about these things, about how they drilled with big old bits. And about the salt water tank at the top. He looked at the top. It had cracked on impact, and the great salt water tower was bleeding out over all of their yards. That, along with the rain, was turning the ditches and divots to estuaries. Some of the neighbor kids ran out and played in it.

But all Remmy could think about was that salt. All of that salt. Right on top of their wells. He stared at it, his eyes staring at the eye in the tower where the… well it might as well have been where the poison came out.

Beth cried out behind him. He turned and found her clutching her belly and leaking some water of her own, the cramps of early labor scared into her from the crashing and banging about.

He swore.

"Don't swear!" she yelled. "This is a blessing!"

"Not in the rain and the wet!"

She screamed. Oh, how she cried out from the pain.

The neighbor ladies came. Beth went inside with them. Remmy fretted and occupied himself with the boys. They went over to the tower to try to stop the bleeding, but the crack widened, rimmed with white, and the water kept on coming. Some of the younger men actually tried to put their hands on the thing, which was about like trying to stop Niagara with your shoulder or some such stupidity.

While they were working over in Joe's yard, they found Remmy's white metal table and chairs in a hedgerow, now dirty from all the mud and sewage that'd slung about in the wind.

"New chairs!" said Joe.

"Like hell!"

"Oh come on, Remmy, finders keepers. Those are nice lawn chairs."

"Well in that case," Remmy said, "I just found this house sitting out here after the storm." He pointed to Joe's house.

"Oh, I was only joking."

"Well joke these over into my yard with me, will ya?"

They hauled them.

"You know," Joe said as they walked back into Remmy's yard, "I heard they got the name *derrick* from a hangman in England."

"Come on, now."

"That's what I heard. Cause they call cranes derricks too. Big towers to hang things from, you know, like a hangman and his noose. Sticking out over to stick something down into a hole like that."

"I'll be damned. Derrick the hangman. It makes sense, I guess."

They never did get the hole plugged. Not with wood. Not with clay or plaster or the pig carcass one of the boys tried. Where he got it, nobody asked because some things just were. After several hours of this, the women came back out and told him he had a son. He ran in to find Beth there, tired and smiling, in the middle of more blood than he was comfortable seeing. The blood there reminded him of a full bottle of ketchup he'd watched get accidentally shattered while he was flirting with some girl at a diner as a teenager, tomato blood scattering all over. But he smiled all the same because he had a son. He named him Tobias—Toby.

Three weeks later, Toby died of pneumonia. It must have been the rain and the wet.

That was the year the Army Corps of Engineers started damming the Kaskaskia River to make Carlyle Lake. Through eminent domain, the government bought out house after house of the townships in the river valley. They capped sixty-nine oil wells.

They exhumed some six hundred graves, some of them undoubtedly holding the remains of ancestors of Native Americans as well as the remains of babies.

WILSON REMUS

1959

HAD THAT BOY who tried to stuff the pig carcass into the hole in that fallen oil derrick succeeded, had the neighbor men succeeded in plugging it with that clay or their shoulders the day heaven spat out that great black tower in the tornado, it would have saved Remmy a good stroke of trouble.

The first thing that oily recycled saltwater from its top tank did was get into the fishing lake nearby—the end point of all the ditches and culverts. Water seeks the lower place, you see, and this explains all rivers and ditches and salt water runoffs from oil derricks that fall from heaven. The fish in the lake died within the week. Then the turtles and snakes followed — the ones that didn't leave, the ones that stuck around. The men who ate the fish got sick. The merry men took to calling the lake The Dead Sea.

Remmy didn't eat the fish. Wouldn't let Beth eat the fish. In the year that followed the whole tornado event,

Beth got pregnant again and had what everyone called their first child — a daughter named Marionette who was born in the oil fields when Beth went to bring John David a lunchbox. Remmy didn't curse his dad but he did curse: he was afraid of another incident like Toby's pneumonia. He didn't want his second child to die a week into life and every time he heard anyone call Marionette his first child, he'd get somber and say, "I had a son. His name was Toby," and he'd go off in the corner and drink or go outside and look up at the night sky and pray for a shooting star. He did even when he got old, and it was a sad thing to see.

But he couldn't have that rest right now, looking as he was at the land and the indented spot where the oil derrick had sat on their front yards. Not that Texarco had hauled it off. Texarco didn't settle on the damage done and didn't even seem to want their derrick back. No, what had happened after it had drained out all of its nastiness and no one had come to claim it was this:

Remmy'd rallied all them carpenters together and he'd gotten ahold of Pete Taylor and some others too and they all sat around waiting for him to speak. Pete Taylor nodded at Remmy once Ryan and Sinclair and Bullhorn and all of them'd shown up. Remmy told them that if no one was going to claim it then they were gonna claim all that metal for themselves, and to get their wrenches and flatbeds and whatnot and they'd work all night if they had too.

Well they did, they worked all of them — all experts in handmade demolition — tearing that oil derrick down into the braces and bolts and things all night long. And Remmy worked up a deal where they'd all buy a shed out west of Bellhammer, closer to Carlyle where he had half a mind to move. Well it wasn't really a shed, it was a warehouse, but they put all of those parts of that tower in there and built up the nicest shelves any warehouse had, repainted them a darker shade of black, and stored all their building supplies and tools in there for a good long while until they'd figure out something better to do with it.

After the shed and that crummy old house, Remmy'd spent his whole savings fund, but that's what they'd done with the derrick anyways and all the men and all their families swore themselves to a solemn oath of secrecy, and since all of them paid for the warehouse, all of them had red on their hands, so to speak. That's how it worked in those days: there were some things you didn't keep no record of because it wasn't worth telling the whole world what you knew more than it was worth watching over the few who were in on it. For instance if you knew the hiding place for the bones of The Great Black Tower.

Remmy thought on this as they hauled it off, sitting drinking a Miller from the comfort of the white metal lawn chair he'd retrieved from Joe's hedgerow after the storm. He'd cleaned it off with the hosepipe.

He couldn't rest from his labors in the company of those he loved. Not yet. He'd have to settle for working in the company of those he loved. And they needed a well, badly. He'd been hauling water from Carlyle every week after the derrick, but it was getting old, so he decided to go to Jim Johnstone, the main boss for that region of Texarco's oil fields.

Remmy walked into this guy's office. It was one of those cheap standing buildings men like that have erected out in the middle of a big project rather than some fancy office building. His fancy office building was back at Texarco's HQ. This was his war tent for the battlefield. It would be left behind like bones of his fallen foes. They had one of those chest-high receiving counters made of speckled white particle board. And also faux-wood panel walls, just to give it a touch of class, Remmy guessed.

"Good morning James," he said. He had decided not to call him 'Jim.'

"Good morning Remmy. How's your father?"

"Working the fields for all I know."

"Mmm." Jim didn't bother to look up at Remmy, just smoked his corncob pipe. Had probably forty of them, one for every pipe tobacco he smoked, treating them like they'd been hewn and whittled by Sacajawea herself, even though they were nothing but stupid old smelly corncob pipes.

Remmy said, "James..."

James snapped awake and looked up at Wilson Remus Broganer.

"That oil derrick landed on my yard."

"Oh do not tell me it's still there," James said. "I told those—"

"No no." He thought about lying about it, but the Good Lord told him, "Don't you lie to him, Remmy, or you'll be as bad as him."

So Remmy didn't say anything more than, "No, the derrick's gone."

"Oh," Jim said. "Well then what's the problem?"

"There was a hole busted into it in the storm."

"As it's our property, you need not be worrying about any damage done to the derrick."

Remmy laughed. "That's sweet of you, James Johnstone.

You realize that derrick leaked that oily saltwater all over our lands?"

"You lived off that pond?"

"It's a lake."

"You live off of the fish in it?"

"Well, no."

"What's the problem?" The man leaned back in his chair. "I hate fishing anyways. You like to fish?"

"Once in awhile, not all that often."

"Well then here's something we can hate together from the comfort of our chairs. Have a smoke with me, Remmy."

"James, I don't care about fish and lakes, though I suppose I should. I care right now about the truth of the matter which is this: your oil derrick ruined and contaminated and adulterated and… and befouled our water supply. All our wells." Remmy was proud of the words he'd come up with. He'd always read more than Daddy John and tried his best to write down new words in the back of all them books.

James was still nodding.

"Well?"

"I don't know what to tell you, Remmy, I'm sorry for your loss, I guess."

"Jim, I want you to pay to put in another well at my house."

"Now Remmy, that's just crazy. That's just crazy asking us to pay for the aftermath of an act of God."

"The Good Lord didn't have nothing to do with the loss of my son, and he did not have anything to do with building no black tower that sticks out its rod to rape the earth."

"Now Remmy, that's not a very nice way to talk about the company that gives your daddy a pension and gives you a chance to build houses for people who keep flocking to Southern Illinois, is it?"

"It's pronounced Illinois."

"Well however it's sounded out, is that a nice way to talk to your employer?"

"Maybe it is and maybe it ain't."

"I think a little bit of gratitude is in order."

"All right then, thank you for giving me, in a roundabout way, a chance to work about a years's worth of income. But I just bought a house from you that's several years worth of income, and without a well, that means you took more than you gave."

James didn't like that. "Terry?"

His secretary perked up.

"Call Brooks and throw this man out. I am not paying to dig him a well."

"I'll show myself out," Remmy said, and he did, and went home.

By that point, everything around The Dead Sea had died — the trees, the grass, and every once in awhile a small rodent like a brown squirrel or a chipmunk. Some of the neighbor kids found a dead badger at one point and carved out its claws and turned them into weaponry.

Remmy asked his father how to drill a well and found out some. He asked some of the boys how to do it, and found out more. He asked this hermit widower farmer out on the edges of the county — guy the kids called a wizard and storyweaver and wombrover who witched wells – and found some stuff he wasn't sure he trusted, but you never know. He read up in the big city library in town. He even called the University of Illinois and they sent him some papers on how to lay bricks up to keep ground water from spilling over, how to make a concrete lid for it, how to protect his own water asset. Their term.

He did it all. Used every last bit of sense he'd found. He witched it and then he borrowed an auger from Kipsy, the guy that normally did these things for people. Kipsy had a couple and said it was okay for Remmy to use it as long as he didn't break it. After that, Remmy rounded up some nice redbrick — a big old pile of it that he stacked neatly next to the house when the time came to put the finishing touches on his — Remmy Broganer's — well.

With that auger, you know, you couldn't start out with a twenty-foot-tall drill bit, you know, else how would you get the leverage to put it in the ground? No, you started with one extension rod, and you could drill down two feet deep. Remmy had a long black power cord stretching from the house like a blacksnake come to watch him, and he leaned over and drilled that two feet. And when he got the two feet and wanted to go deeper, he had to pull it out, pick up a piece of pipe, and add it to extend the drill bit and then drill again a little deeper. Then you'd pull it out again, add another two-foot extender and drill a little deeper. And he did this all into the evening after work, down and down, adding extensions and extensions. And when he hit water it came right up the hole. He tasted it.

Salt water.

"Awwww shit," he said. "Aw shitty shit shit." He pulled out the extensions one at a time, wrapped up the auger, recoiled the blacksnake powercord and went in to bed.

"How did it go honey?"

"I hit saltwater," he said.

"Shucks," she said.

"Yeah, that too."

The next day, after wearing out his arm hammering studs together, he went back to the auger at the opposite end of the yard. He dug down two, four, eight, sixteen, almost thirty feet and hit water again that night and he smiled. He tasted it and tasted its brackishness and swore some more and would have thrown the auger if it wasn't Kipsy's tool. Taking it apart had a soothing effect on him, or at least created a little buffer where he could calm down before going in to Beth. He stacked it up and coiled the cord and went inside.

"Did you get it?" she asked. Marionette cooed over in the corner somewhere.

"I got a hole, if that's what you mean."

"No water?"

"Oh I found water too."

"Oh good!"

"Saltwater."

"Oh Remmy, I'm sorry."

"Love you," he said and kissed her forehead. He walked over and tickled Marionette, who giggled and giggled. As far as I know, that's the last time he tickled her — got too busy to think about it.

He hauled water in the morning and a couple of days later he tried again for a third well, which turned out to be salt. And a fourth. And a fifth.

After six wells, Remmy gave up on the auger and spent a day giving it back to Kipsy.

"Go well?" Kipsy asked him.

He laughed a dark laugh. "Funny old turn of phrase, that."

"Did it?"

"No, but I thank you all the same."

"Need help with it? I can show you how to use it."

Remmy was not insulted because Kipsy didn't mean offense, but it still hurt a little bit, the truth of it. "I worked it okay, thanks Kipsy. Just hit saltwater every time."

"You're pulling my leg."

"Wish that I was."

"In that whole big yard of yours?"

"So far," he said. "So far."

"What are you going to do, Remmy?"

"I'm gonna try a dug well next week."

Putting in a dug well with a short-handled shovel was not the best way to go about it, but it was the only way he had at the time and all he had energy for, which was a fine joke on him since the short-handled shovel ended up taking more energy from him than had he gone hunting for the other kind. But anyways he went to digging and had himself a hole six foot across and he was pulling up the dirt and the mud and slinging it up and up and went through a softer and softer soil, which he did not expect, but he kept going deeper, digging down into the soul of the place to find if it too had been corrupted.

At around fourteen foot deep, he drove his shovel down and got into the sand. He got excited then, as if he'd dug through from Illinois and ended up on the riverbanks of Korea.

He went into the sand all the way, laying up shovelful after shovelful until it piled up outside the hole. He threw that shovel one more time into the sand and the earth swallowed it clear up to the handle. He'd hit good water, he felt sure of it.

Boy, did he ever hit it.

It came up quicker than oil, quicker than anything his father John David had ever seen working in the fields and the derricks. Gushing.

He speared that short-handled shovel up and out of the well he'd dug and he started scrambling up the wall of the well and ran over to the pile of bricks and he loaded them up in that old wheelbarrow and brought them over and dumped them out and mixed him up some mortar as quick as he could and started laying the brick and laying the brick in a great big old circle, the well from every storybook, its cup runneth over. And no matter how fast he laid the brick to line the well, he couldn't beat the flow of the water, the water line beating the height of the liner over and again, so he called up one of Texarco's tank trunks to put its hose in there, which they were happy to do, taking free water, and that tank truck pumped it out as he laid in brick.

But still he couldn't lay brick fast enough. He laid about as best as he could in the water there until he had it laid up six feet off the ground just like the University of Illinois said to. He did it just so and knew he would not get surface water in there. He had ten feet of well, he did. And he set some boards down in there. He poured a concrete top and fitted it to the top. Then he put a stepping box next to it. He bought a pipe big enough to get the stuff out, you know, to get the pump out of the ground so he could get to the water when he wanted.

He got it really nice. And after he'd done all of that,

he took a glass and scooped some out and tasted it and it quenched his thirst. Then he scooped himself another glass and took it on into the sanitation department.

"Morning Remmy."

Remmy was smiling. "Morning Tom."

"What you got for me, another test?"

"I do, Tom. I did a dug well this time, built it up really nice."

"Oh that's good, Remmy, that's so good. I was worried about you." He took the glass and ran the tester. "How's your daughter?"

"Getting big. Getting pretty. I'll be beating off the boys with a four-by-four soon enough."

"Well good luck with that." Tom looked down at the test and tsked. "That's got enough salt in it, you could float an egg."

"HORSESHIT AND APPLEBUTTER!" Remmy loaded what he had with the test up into his car, and he drove back out to the oil fields. He smelled that smell of a sewage or sulfur line filling up the land. Silt and sod. Good soil is really just good shit that's had enough time to mellow out, and in a land covered in rivers and held up by oil wells, the smell of sulfuric shit's high and low and near and far and wide in an oil field. Remmy walked right up to that makeshift Texarco office and opened the door so fast — it was an outward-swinging door — that it hit off the outside and bounced back to slam shut behind him. He cleared that chest-high countertop and jumped over the desk, too, and had that lazy asshole Jim Johnstone up by the collar before the man could finish his shout.

"Leave him alone. Don't be starting a fight," The Good Lord said to him.

"I want a well," he said to The Good Lord.

Jim answered something tepid, but Remmy wasn't listening to that stupid sonofabitch. He was listening to The Good Lord.

"I will give you Water," The Good Lord said to him.

"Okay, then," Remmy said.

Jim Johnstone thought Remmy had said Okay to him, had agreed to whatever pissedpants deal he'd struck up in the moment. Remmy left. But he must have scared Jim anyways, which was not a bad thing on general principles.

Remmy went into town and filled out paperwork to start building a house in Carlyle on the lake, near water. And in the woods. Like Robin Hood. Perfect place to finally rest in the company of those he loved. He came back to his house that night and talked to some of the boys about his plan and they liked it and got finally why he'd wanted The Shed where he'd wanted it. But they didn't like his tone about The Woods. The plan got back to Jim by the next day, and then Jim's reaction got back to Remmy:

"Well that's a good thing," Jim had said to his troops and his boss, "because I think I was just about ready to settle with him. That man had blood in his eyes. I would have given him anything. I'm glad he's out of here — sooner the better."

When he heard this, Remmy said to his neighbors, "Well I probably should have gotten an attorney, but it's over and done with now, and we'll have a lakehouse and all the living water we want."

Beth couldn't be happier, of course. She could buy more tables.

That wasn't before he washed the car twice with that salt water, though. Twice was all it took to corrode that chrome bumper to rust. Imagine had they drunk it.

WILSON REMUS

1960

AT THE TIME, he was building freeze boxes for Texarco. Made out of cement asbestos board. Build a wooden frame so that it wouldn't rot on the inside — the food — and nail asbestos on the outside of that box. Because you put insulation in on either side, it'd preserve. It was not very good for you to breathe, though. That was the starting of his COPD. A fire in later years made it worse.

Meanwhile, Remmy'd be siding his own house at five in the morning before he went to work. He'd side, then build the other Tulsa Rig and Reel houses, then come back and hammer a little more right after supper before he called it quits. See he'd sold the shack and reinvested in that dumpy Texarco house that the oil derrick fell on top of and then sold that one for this. Texarco tried to take their bridge back again when he sold it, and he paid the new owner — who hated the bridge cause he didn't respect good woodworking— $20 just to keep the bridge and write it into the contract. Texarco was so mad. And Remmy wasn't happy, but getting taken for twenty dollars and paying forty to stand up for yourself is two different things. So he took that house snowball — the chunk of change that still contained the seed of CAMELOT, MY MERRY MEN, AND OUR PRANK fund — and he rolled it into his new house on the lake, where he'd hammer in twilight and dusk alike.

That year he read *Doctor Zhivago* and found himself moved by the longing for steady home life, by Zhivago weeping on his mother's grave. I came across his old copy these days and some of the pages are warped from teardrops.

He'd save a little and go on a vacation with Beth and the baby for a few days — at one point to Santa Claus Land for a day or two, like he'd always wanted to do with his daddy — and he played in the living water with Marionette on the lazy river and squirt water at Beth who was always tanning with one of them trifold aluminum foil reflectors, and she'd get mad at him for getting her wet, and he'd blame it on Mary, who'd giggle. He tried to get Beth to ride a coaster and she didn't want to and Marionette wanted to, but he couldn't get her to ride on account of the meter stick beside the ride. They'd rub Saint Nick, patron saint of pawnshops, for good luck and then he'd come back home and bang on his own house a little more. It already had water from the reservoir, you see, and banging around like that was much better than spending your days using an auger to poke into every nook and cranny of the land looking for water.

After the vacation and while he was banging around one Saturday, Beth was inside making curtains. Beth could sew something fierce. She could have sewn the regalia on Sir Lancelot's horse blanket. She could have sewn the gowns the queens wore and mended the hats of generals. She, like Remmy, poured her very soul into her craft and she got damn good at the things she did because of it.

She was sewing curtains when her labor pains started. She'd been pregnant most of the year with a baby, and now he was coming. She moved the drapes out of the way and realized that in the pain she'd slipped and sewn her gown to the curtains, the looser fabric she'd needed to wear on account of her swollen belly'd gotten caught in them.

"Remmy!" she called. "Come quick!"

He didn't come.

"Remmy!"

She could hear him hammering away outside. So she got her scissors and cut at the stitches, missed, cut a nice round hole in her new blouse, tsked at herself, got up, and walked outside with her belly sticking out that hole.

"Wilson Remus Broganer, get the car!"

Remmy looked at her with his hammer mid-swing. "Why?"

"Because you're gonna have a baby in your lawn if you don't hurry your ass up!"

"Ah." He dropped his hammer claw-first so that it stuck in the earth as would the hammer of Thor and he got in the car and asked her as he was starting it what happened to her blouse and how'd it get to where her belly looked like the head of Friar Tuck?

She growled at him.

"Okay," he said. The only thing he hammered on then was the gas cause he wasn't about to continue the tradition of babies being born in barns and wagons and oil fields. Remmy succeeded as the first American Broganer to get his pregnant and laboring wife to the hospital, or at least to the nuns up in Greenville.

First thing that head nun said was, "Welcome, Mrs. Broganer. How'd your blouse get to looking like the head of a friar?"

And Beth growled at that old nun too. Takes a lot to scare a nun, but Beth in labor was a lot. They stared each other down in the way that only two women can do.

"Where's your daughter Mary?" the nun finally asked.

"Oh shit!" Beth shouted.

"Is it a curse?" Remmy asked.

Beth shouted a string of cursewords, some of which I refuse to write down and others I'd never heard of and had to look up and still others I'm pretty sure she just made up or found in some foreign tongue. Or maybe came from some other world, some tongue yet to be translated into Narrative.

"Think she needs an injection of something calming, Doc," Remmy said.

Bren was born quick as a pistol shot. Brennus Patrick Broganer. Up until then, Remmy'd thought his days of pranking and being pranked had died with his Grandad Patrick.

Then came Bren.

WILSON REMUS

1961

Tulsa Rig and Reel called them in for their checks after a good string of houses, called in both Daddy John David and Wilson Remus. The big boss, Tommy, passed the checks across and said, "Tomorrow morning your job is in Joplin, Missouri. We'll hire you a crew down there."

"In where's it?" Remmy asked.

"Joplin, Missouri."

The only encounter Remmy'd ever had with Joplin, Galena, or any of them towns was from the lead in his head that'd once coated Texarco parts, lead that'd come from the Four States area and through a lack of policing had come into the water momma Midge'd drunk while pregnant with him. All Remmy knew is that it was far away'm Southern Illinois.

"We just moved in," Remmy said.

"I know," Tommy said. "But you can sell your house and then we'll buy you a nice new one there out west."

Remmy snorted, and with much struggle on his part refrained from saying how stupid that sounded.

John David wavered for a bit.

Remmy walked his dad outside. "You're not thinking about it are you?"

"Need the work, Remmy."

"You could always start another farm."

"At my age?" John David laughed. "Why you always living in the past, boy?"

"It's not the past I'm living in, Daddy John," Remmy said. "It's some bits of good the past once had that the present don't have no more no matter how big and neat all these machines get. I'm after that goodness. Hell bent to find it again."

Daddy John looked at him with sad eyes.

Remmy looked back.

"Hell-bent don't sound good to me," John David said.

"Why'd you ever sell our farm?"

Those eyes again. "You don't get it, Remmy. You never did. You never will."

"So help me, Daddy John. Help me see with your eyes."

"Shit, Remmy, don't you know you could buy land here for three dollars an acre because it wasn't any good? How the hell you think Texarco got so much of it so quick? You couldn't raise dandelions it was in such bad shape. Them farmers only just started buying them tractors, and now it's starting to produce. But even still they gotta drag them tractors through the mud sleds when they get stuck. I was tickled to sell my land at the time."

"Not me," Remmy said. "Soil black as that. Our land was full of milk and honey."

"It was that," John David said. "It's just, with the tools I had I couldn't till deep enough to get the milk and honey out. I'm just a regular old worker bee."

Remmy saw that his Daddy was determined not to let him see no tears. And he respected that.

They watched one of those big old earth movers go on by. They sat watching as it scaled the hills of dirt it had spat out after chewing up the land, climbing up the refuse it'd made like a giant wyrm.

"Whatcha really think about Joplin?" Remmy asked.

"I think building houses is the only thing left for me to do," John David said.

"Then let's build houses right here, Daddy John."

"For who? Tulsa Rig and Reel wants us in Joplin. There ain't no other. Who you gonna work for?"

"For us," Remmy said.

"You want to start your own business?"

"It's that or work for the bank."

John David laughed. "You try the bank first fore you do something crazy like try to bear up the weight of Little Egypt on those boney Broganer shoulders of yours. I'll do some odd jobs while you figure yourself out. And when you're ready, we'll pack up the whole Broganer clan and move to Joplin."

Remmy did just that. He went and worked as a teller at Centralia at a bank owned by Beth's old boss, and hung onto his first check from them just in case — he had enough money from Tulsa Rig and Reel that he didn't need that first check to carry him through. The boss was a good old boy and gave him a chance to prepare to be a Certified Public

Accountant, which'd been his other idea when he'd graduated high school and hadn't taken physics.

But the further he got into that diploma of his, and the longer he spent staring at other people's money, Remmy realized that the only numbers he cared about were angles and lengths. Fractions, not decimals. Sines, not cents. Compound labor, not compound interest. He tried to get along with the staff, and they were nice enough, that wasn't it. He tried flirting with the other girls there, but after awhile it was either cheat on his wife or realize that he should be closer to her and not flirting down in Centralia with a bunch of strange women he didn't give a shit about cause they hadn't borne his kids and lived in a shitty shack with no well water and made him feel strong when Texarco made him feel weak, so he stopped all that nonsense. Then nothing was left but the boredom.

But the boredom bored a hole right through his head, which gave him plenty of room to think. Really is quite useful, being bored. That's the thing city folk forget about the country and it's one of the reasons old Emerson said the city's recruited from the country, boredom. That and the effect of birth rate and small businesses on the overall economy, didn't Warren Buffett say so? But anyways it's useful being bored:

The way he saw it, he'd pretty much been starting companies as long as he could put money in a can. He started the ice cube business. He started the Barbed Wire Telephone Company. He'd start the Spur Cola auction house some day when he finally could afford to hire a team of treasure hunters and go on that grand old treasure hunt to dig all them bottles back up.

Hell, he'd married a banker. She could run the books for him, and then it'd be nothing but the building and the bidding. Well he told them when he'd be quitting, worked all the way up until the end of the last pay period as many hours he could. He still had his first check. Cashed two checks that last week — that's what he had left to start his business — and walked up to John David with the money. "We're officially starting the Bellhammer Construction Company."

"We are?"

"Yes sir."

"Why Bellhammer? Why not Broganer?"

"Cause Broganer means shoe maker, and we're more than cobblers. Also cause sometimes hammers can do more than stitch a house together. Sometimes hammers can sing a song. Like in a bell."

John David laughed. "Okay, son."

"And cause if nothing else it's the name of our town, the one we plan on building bigger."

"I said okay, son. Let's give her a go."

The go they gave her was building another house for Old Kergan on the corner of Illinois and Shelbyville streets.

Fixed up the roof and the. Real nice carpet — Remmy took off his boots just to curl his toes on the stuff. Made it as good and big as they could for the price because this first house was going to be the best ad they could get and get paid to do. And Remmy exercised his love of angles in making the roofline interesting and adding a bumpout picture window and a couple of artfully placed gables.

People was coming to town to stare at it. Think of how they could scrounge up money to buy one of their own or gossip about all the people in town that might have enough money to get one for themselves, so custom built compared to the cookie-cutter minimal traditionals they'd been putting up for Tulsa Rig and Reel. The people all took down their names and got excited about having nice houses and moving down to the land of Texarco and sweet corn. So many wanting to talk to Remmy and John David.

And then the biggest snowstorm we had in years hit the Bellhammer Construction Company. Four feet high on the flats, and drifts beyond believing. These people went back to where they came from. Remmy lost track of all of them. What would you do when you had no money to build another?

What could you do but wait and pray?

It was eating breakfast, lunch, and dinner in the evening. It was playing with the kids, with the two-year-old Marionette and with one-year-old Bren. And Remmy actually got John David to come over one day out of thirty to play with his own grandkids, but Daddy John didn't stay long. "I like 'em better when they start doing things," he said and he'd looked at Remmy with a sort of longing. "Come with me," he said. "Come do some odd jobs."

Remmy said, "I think I'm best suited to wait here with my family and enjoy the rest God gave me."

John David looked out at him with hollowness, like one of Peter Pan's lost boys or one of the older Merry Men that Robin Hood hadn't met until he was already there in the woods. "Alrighty then, let me know when it sells."

Remmy'd wanted his daddy to stay around. In fact, he took the kids over to Daddy John's house as often as he and Beth could afford in that magic Chevy, still purring along after all these years thanks to his keeping the oil clean. But Momma Midge'd always say, "He's off in the fields" or "He's repairing Janine's porch," or what have you. She never watched Bren and only watched Marionette once — more on that in a minute.

Six months later Remmy sold old Kergan's House.

Daddy John said, "Oh, thank God. I thought we'd starve."

Remmy said, "Still better than working for the bank."

John David said, "Son, since you've never gone without a meal, let me tell you: no it ain't." And he left it at that because he was staring off somewhere, back to a more depressed time.

They built another house and those bankers gave him $1,000 — but he had to wait to get it. No, that wasn't it. It was a bit more. $1,200, and they financed the rest. Yeah, that's it. So Remmy had to wait for them to pay so much a month to the bank before he got it back. He was living on payments in that time, almost like he was retired or a bank himself, which is much the same thing, everything and grandkids considered. Speck houses. He thought they named it that because eventually you'd get to where you'd have yours all over everywhere like specks on the map. That was his goal, specks on the map.

Wasn't until he was desperate for that next paycheck and ran it as soon as it'd come that he found out what the word really meant.

"No, no," the banker said. "S. P. E. C., Remmy.. Like speculation. Cause you're speculating as to whether or not it'll get built."

"Well that's a dumb way to go," Remmy said.

"Yeah, but a quick one if you're not careful."

"Uh huh," Remmy said. "Debt can get you dead."

"Uh huh," the banker said. "That's why I don't owe people. I have people owe me."

"You're in the right business."

"I don't know that I'd call it good employment, but I'd call it gainful."

After that, Remmy thought the best thing to do wasn't building to sell later. He decided to only build what he'd already sold.

Beth worked as a teller for that banker still, and Gwen would watch Bren and Mary and teach them how to knit and sew, from which Bren later started sewing booby traps up all over the house as soon as he could climb. Beth working for him's why the banker'd been honest with Remmy: he liked Beth and liked Remmy by association and had even invited them to a party that had a giant shrimp cocktail, of all things. That was the first time Remmy'd ever had shellfish. Or really any fish that wasn't cat or bass or something caught in the dam like striper or sauger. Them sauger's big but easy catching if you feed 'em your own flesh and blood. He'd sometimes coat chicken livers and baitfish with his nail filings and dead skin and prick his finger over them. That was the trick: feed 'em your own flesh and blood. So yeah, that shrimp cocktail was something: he wondered how they'd caught 'em all. And where.

They couldn't sell that spec house, so Remmy tried to sell HIS house on leased ground. He was leasing it there on the lake, you see, and it just don't work so well when someone's gotta pay a mortgage and the rent to two different men. Finally that spec house sold, thank The Good Lord, and that was the last time he ever speculated about whether to build a house he hadn't already sold to somebody.

But he still thought he could get his houses everywhere like specks on a map.

He took some of the money from the spec and bought a pretty tapped-out Dodge, the miter box, and a torque wrench for $135. Good Lord, he hated Dodges. His Grandad Patrick'd been right to dodge those purchases. Remmy put The Magic Chevy in The Shed for when Bren got older and started fixing up that Dodge with Marionette in mind. Them old ones was hard to start. He bought two in a row.

"I was a dummy," he'd say one day.

That stupid car didn't start when it was really hot in the summer time. It also didn't start when it was really cold in the winter time. In fact, the only time it did start was when the crops themselves were starting in the spring, and only then when there wasn't no rain, cause the humidity made it didn't start. It got over 180k miles back when cars didn't cross 100,000 that often. Near the end, the motor got so loose that if you drove it half a mile, oil leaked past the rings and fouled the plugs and had to be tow started, but after two or three miles it warmed up enough to burn off that oil and started again. Only way to fix it was to buy a motor that probably wasn't worth rebuilding.

When they first got it, they wanted to get it started, so Remmy had Beth pull the Dodge with The Magic Chevy using a twenty-foot chain. Beth's in the Magic Chevy. Remmy's in that beat up old Dodge. Marionette's being babysat by Momma Midge for the first and last time. Midge said that nannying gave her a sick headache and never did it again. Well Beth in the Magic Chevy's pulling Remmy in the beat up old Dodge and she hit the brakes so fast that baby Bren — who she had up sunbathing in the backseat window — fell down and out onto the backseat and started to giggle. Well she parked the car and brought Bren up front, who started to cry cause he liked it up high in the window like that.

On the next pull, Remmy got the Dodge started and he hit on the gas right when she hit on the brakes, and he was gunning it towards that backseat window. He saw her. And wanted to be able to get stopped fore he hit her, so he hammered on the brakes and squalled them tires and the chain went loost and he went off the side of the road in the ditch. But he kept it rolling and got back up on the road in front of her and he stopped and the Dodge died.

Beth put baby Bren back up in the window where he sighed there in the sun, and they went off again to get the Dodge started and she hammered on the brakes again at the next junction and Bren fell down again giggling, and Remmy got The Dodge started but almost hit her again and swerved again and so they went like two train cars with slack in the coupling until they made it home.

And then the next morning, the Dodge didn't start.

WILSON REMUS

1962

"WE HAVE ANOTHER Cuban crisis, we'll all need them," Norm said.

"Well call me up," Remmy said, "and we'll hook you up."

"I bet you've been building all kinds of stuff cause of the bomb," Dale said.

"Dale, by God if you don't stop moving—" Norm started and then he got this wicked grin on his face.

"Oh yeah," Remmy said. "We built bomb shelters for the real estate dealer in town. He had a nice house and a baby tractor. You need that much more room. He wanted a bomb shelter beneath it all."

"Well how'd you do it?" Norm asked, shaving Dale.

"Just dug it deeper," Remmy said. "We cut a hole through the concrete basement, set the footings. Then we got a concrete drill, put in beams and feathers and this fin thing went into it. We drilled a feather and you could take the whole chunk out."

"The floor?"

"Yup. Then you had a bomb shelter underneath. We built twelve-inch-walls, double reinforcement, triple reinforced that, thick on top of it, dug it deeper — about twelve inches more than the joist was mounted to — and matched up."

"You do a lot of those?" Hank asked.

"Yup. Just in case," Remmy said. "Thank The Good Lord we didn't get hit in October. Could've been bad, but we have them shelters now."

"We'll call you," Hank said. "Everyone I know."

Remmy nodded.

Dale got up and looked in the mirror. He was as bald as a trailer hitch — not a strand of hair on his head other than eyebrows. "What the hell, Norman! I told you I wanted a haircut like Frank Sinatra in *Seargents 3*! Not Telly Savalis! Frank Sinatra don't look like this!"

"He would if he came here," Norm said. "Specially if he moved around as much as you. That'll be two dollars."

Dale gave him the money and asked for change.

Norm put it in his pocket, went over in the corner of his shop by the trashcan, and started pissing against the corner wall.

"The hell?" Remmy asked. "What are you doing to your shop, Norm?"

"My lease is up. I'm leaving in two weeks." He went to the change drawer and started getting Dale's change.

Remmy started laughing.

Norm turned around to the chair and Dale wasn't there.

Dale was in that same corner, drawers dropped, taking a shit on Norm's floor.

"The hell?!" Norm shouted.

"Well I'm leaving right now!" Dale said and ran out the door.

Remmy laughed and laughed and got himself a haircut anyways, smell or no.

"Guess he didn't want his change," Norm said.

"Guess not," Remmy said. "Whatever you do, Norman, don't you dare make me look like Frank Sinatra."

Norm said, "Not even a dancing Rat Pack moves as much as Dale."

Later that week, on Sunday in the cold, Remmy sat on Daddy John's back porch, looking out at his father's big yard and the train tracks in the distance, and that big old stone in the middle of the yard nearer to him, listening to the children play inside, feeling the worn wood of the arms of that old rocking chair, smelling the snow that'd just barely glazed the ground. He liked feeling alone when people were nearby — he could always call on them if he needed anything and go in to help them if they needed him. But something about being by yourself just a few feet from people you love gave Remmy a good stroke of comfort.

A flash burst in the sky.

He watched it.

It came closer and burned brighter and did not go out.

It cut a hole clean through cloud and came right at him from the sky, burning and growing, and he could see the dark core of it then and he thought, "Good Lord, the Cubans finally did it. They launched a missile."

"No," The Good Lord said. "That was me."

When it hit it threw up dirt like a tiller the size of a

derrick would have thrown, cutting a giant ditch in the ground. It plowed straight to that stone in the middle of the yard and the bigger rock stopped it.

It sat smoldering.

"What the hell was that?" someone called out from inside the house.

Remmy walked out to it and stared at that burning meteorite, its surface pocked like the old acne-scarred face of Mr. Tolliver. "Why in the hell have you brought all these plagues on me, Good Lord? First The Black Tower full of salt water thrown by wind and now this fire and brimstone?"

"You're not paying attention, Remmy," The Good Lord said. "Open your eyes, boy."

So Remmy started watching and waiting.

A few days later he said, "Good Lord, I'm going to build the Southern Illinois Museum of Space and charge a dollar for everyone that wants to see the space rock."

"No, Remmy, you will not," The Good Lord said.

"Okay, good, I'm glad you say so cause I don't know nothing about space and don't like it neither."

"Open your eyes, boy. Take and eat."

So Remmy opened them and waited. Meanwhile he called up Pete Taylor and the carpenter boys — his Merry Men — and they all weighed it. It weighed exactly three hundred and thirty three pounds. Then they loaded it up in the back of The Shed and hid it under the world's largest tarp. Pete Taylor got to giggling as they did it shaking his head. Somewhere in there, Daddy John hit it with a hammer and it sparked.

"Hell, Remmy," Daddy John said, "That's made out of iron and nickel and not just brimstone."

"I'll sell it for a pretty penny," Remmy said.

"No, Remmy, you will not," The Good Lord said.

"Well, dammit, then what?" Remmy asked the Good Lord.

"Stop trying to make money off of my stone. The answer'll come if you listen and learn and make the time. The answer'll come and pick it right up for you."

WILSON REMUS

1963

"NOW THAT THERE's pretty," Remmy said. "That's a fine appliance right there."

"I don't like the color," she said. "Looks like curdled cream."

"Oh now, don't you start in."

She started in. "You plan to cook on this?"

"Now that's the same thing you said about the table. You can't use that line twice on me."

"That was at the other house. Stuff at the other house don't count here. Don't you be tracking salt into this house."

"You try digging oneteen wells and come back and tell me that don't count!"

"Well it's just curdled cream and I don't like even sweet cream unless it's on a pie or in a cup of coffee and I'm the one cooking on that cream horror."

He sipped in air between his lips. "Look, will it work for what you're cooking? Will it *work*, Bethy?"

"Oh I wouldn't ever complain about it thataways," she said. "It's a striking piece of kitchen equipment, Remmy. It's useful."

"You're just saying that."

"No, really. I've never seen a stove so big. I hardly know what to do with two burners let alone five." She actually sounded excited.

"Well good, cause all of that red sauce and meat grease would mess up a white stove anyways."

"I was thinking black. Or the color of a gun's metal."

"Listen to you," he said. "Who on the wide earth would ever color their stove black or the color of a pistol? You already killed the food — ain't no need in killing it twiced. Cream or white's about the only two colors on offer, you'n go look."

"We coulda taken out a loan."

"I ain't having my bride cooking on spec!"

She didn't say nothing.

He continued. "Cream or some Easter pastel, and I didn't think you'd want to be wearing the same color you're cooking upon during Easter Sunday."

"That's the truth enough," she said.

"So should I hook it up or take it back?"

She smiled a little, her shoulders shrugged, her chin tucked. "Okay, then."

"Okay then." He bent down under there and started squirreling around, hammering and soldering where he needed to, and getting the wires worked out like he was supposed to, and once he got it all hooked up, he went over to it and turned on the burner with the plastic knob

— touching nothing else, mind you — and the burner came right on in that electric stove.

Bethy nodded a curt nod.

He hugged her.

She didn't hug back. At first.

Now the problem was that Remmy was no electrician. None of us Broganers do good when we mess with sparks and shocks and brimstone. For the Bell Hammer Construction company, that was Billy Phipps's job on most of his worksites. So Remmy had no way of knowing that after his soldering job on that cream colored stove, one of his wires kissed the oven door when it closed. And if you made contact with the metal handle of that door while the burners were on, it completed the circuit. The word for this kind of electric problem is a "short." It's what happens when the circuit takes a shortcut through some other device — sometimes a penny, sometimes flammable insulation, sometimes a human body.

He went into the bathroom that night while Beth got to work on dinner, and he had all of the lights in the hall and the bathroom on so that he could see to shave, all lathered up with that barber cream and the badger hair brush he'd bought from Norm. The lights flickered. Then they went out and back on in rapid flow alongside that sickening hum like the buzz of a mile-long bee come to claim its keeper or its queen.

He dashed into the kitchen and he saw his wife clinging to both the handle of the fridge and the handle of the oven, shaking there, 220 volts of electricity flowing right through her. He almost went to tackle her and would have gotten electrocuted too, but it didn't come to that because it was over quickly.

All that electricity ran through her muscles, seizing them up, and so her hand was stuck to the oven. But her weight let her slide down and that broke the contact. Her weight overcame the power of her opposable thumb, even when her thumb had been boosted by the power of 220 volts.

Remmy caught her before she hit the floor. Would have gotten electrocuted to death himself had she not been falling.

He held her and held her and held her and said, "I'm so sorry for being such a quarreling cuss."

She just whimpered a little.

He put his hand through her hair.

And she looked up at him and said, "You fool."

"Can't argue with that," he said.

The next morning he called up Westinghouse with Grandad Patrick's fire in his blood.

They came out to the house like Philip Morris should have done for Grandad.

And out there at the house, after tinkering around with their cream colored, five-burner stove all morning, they said, "It works fine, ain't no short in it."

To which Remmy replied, "By God Almighty, put the meter on it and you'll see!"

The Westinghouse man said, "I don't do anything by God, mister."

"Ain't that the truth," The Good Lord said to Remmy.

"I can see that!"

"Well just be careful and don't let it happen again."

They left.

"I should have got an attorney," Remmy said to Beth.

"That makes twice," she said.

He hooked it up again and then got it hooked up to the wall power. He still hadn't noticed it wasn't grounded. Well experts had just said there wasn't anything wrong with the stove.

Still worried, he took her to Dr. Nesbit, and there wasn't anything the doctor could do. "I can see you've been through quite a shock."

And Beth said, "No shit."

And Remmy said, "I thought you didn't curse, baby."

And she said, "I do: at cursed things."

And he smiled.

"She has to learn to stop getting ahold of that stove," the doctor said.

So she did. She never touched that stove handle again without a dish towel wrapped around her hand.

Remmy wasn't done, though. He'd barely started getting ahold of the issue. He called up those Westinghouse boys again and said, "I have my doctor's bills here, and I want you boys to pay for them."

They said, "What for?"

"For electrocuting my wife as if she was a mass murderer instead of the mother of my children."

"Would this be Remmy?"

"Yessir."

"Let me give this to the manager."

"You do that."

There was a muffled pause and the sound of voices

talking low and wandering as if underwater. Then the rush of air.

"Hello, Remmy?"

"Hello there, with who am I speaking?"

"This is Larry up here. Larry Tiller. Look, Remmy. Decisions like that, to pay a doctor's bill, ain't gonna come from me."

"Thought you's the manager?"

"I am, sir."

"I'd have thought the manager could manage. Who's above you, then?"

"The owner of this notable franchise."

"Who's that?" Remmy asked.

"Why that's Jim Johnstone, the Texarco man."

Remmy bout electrocuted himself just standing there stewing. "I'll call you back, Larry. Thank you for your time."

"Yes sir."

The men hung up mutually, and Remmy wondered what he'd do.

WILSON REMUS

1964

AROUND THAT TIME, Remmy built onto the house. He built a really nice family room with warm wood and brick colors. Really nice place, the family room. But the problem was, that side of the house had a hole in it while still in the building stage on account of the studs. And even though he locked that inner door to protect from people coming into the house proper, even still someone kept coming in through the studs into that half-finished family room. A stealing his tools.

Tools was expensive.

So he bought himself a really cheap doghouse for putting tools in and almost put it back there with the opened up wall and the family room, but it was so muddy back there, he couldn't do that.

He put it in the front yard instead while he was building the addition. Put his tools in there and locked it with that damn combo lock that once made him shit his pants.

He figured if a lock's good enough to make its owner shit his pants, it's at least good enough to make a thief piss theirs.

One of the new people who'd moved into the Dukes and Dames side of the subdivision – not on the sharecropper side — was none other than Jim Johnstone, the Texarco VP and owner of the stove store, the guy they'd once kettle pitchered after they'd gone and cat whipped his elder brother, same guy that as a kid had announced the bombing of Pearl Harbor with that look on his face like he had bad news nobody else knew about and he'd only tell you once you begged him good and long. Sweet as Jim sounded when he talked — or at least as sweet as he tried to sound — when he didn't like something, he did what he pleased to fix it his way. And even though he was a duke, he was also a neighbor, which made it worse. You gotta love your neighbors, like The Good Lord says to do, or else they turn to foes, and nobody wants to live cheek to jowl with their foes, cause then you got double work trying to love your enemies on top of neighbors.

There was an old barn across Edgewood Drive — Edgewood was the neighborhood. Jim didn't like that old barn in *his* neighborhood – and, yes, he really did talk like a mob boss or a gangster. Complained about it all the time to the other dukes. When they played out in the yard next door, he'd complain to them. He'd call them up in the middle of the night in the middle of the work week just to whine about that old barn.

One day, Jim drank too much, and took up two gallon cans of gas. He wanted the neighbor to come over and set that thing on fire. He was shouting out in the yard, "Come on now, Horowitz! Come burn down this barn for me!"

He was shouting and screaming and half the neighborhood came out to look at the fool.

Horowitz took that gas can away. Both of them. And he said, "Go home."

Jim looked out at the rest of them and said nothing.

When he got home, they could all hear Mrs. Johnstone shouting and kicking his butt for being drunk. And if there's one thing more dangerous than a wife beater, it's a man who's been beat by his wife so much he's got no dignity left and looks at the world like it's being shown to him through some funhouse mirror. Man marked up like that'll do anything just to leave his mark.

He came out in the middle of the road and said to Remmy while Remmy was working, "You can't have that shack out here!"

"Dammit, Jim," Remmy said, "I'll move it as soon as I'm done with it, but they steal stuff out of here."

"Oh come on, Remmy," Jim said, "who would steal your tools?"

Remmy growled.

A couple of days later, some other of the rich neighbors from the duke and dame's side fussed.

"I'll move it as soon as I'm done," Remmy said.

Beth said that night while making a pork chop, "It ain't hurting anybody. Leave it 'til you're through."

"Don't you see, Bethy? We're on the sharecropper side of the street."

"What's that got anything to do with it?"

"So rich folk think if they can see something poor, it hurts them. They think if they see a bum, they might become a bum. They think if they get asked for money and get involved, they'll risk their good name and their good fortune."

"Why?"

"All their friends might think they're poor and sick too."

"That's about the stupidest thing I ever heard," Beth said. "The Good Lord was homeless and crucified."

"That's true," The Good Lord said.

"See?" Beth said.

"The Good Lord was made that way by rich folk."

"Also true," The Good Lord said.

Bethy seared the meat.

Remmy sang the Woody Guthrie song about the rich folks laying Jesus in his grave cause he told them to give their goods to the poor.

Two days later, they had a fire in that shed.

Burnt down everything it held that could burn, melted some weak metals, and smoke damaged everything else, that fire. (He'd will those soot-black wrenches to me and my brother.) So Remmy loaded the remnants of the shed up and sold it to somebody else. You might be wondering who'd buy a burnt shed, but that's because you've never been poor in Southern Illinois and seen what a man can do with a half foot of scrap PVC pipe, some leftover WD-40, and a book of matches from the local pub.

And Remmy had to leave his tools out in the open again to be stole by the thief. He had an idea of who'd set fire to the shed, but he didn't have no footprints or fingerprints, and he couldn't hire Dick Tracy, so what could he do except call it a lazy hate that'll burn a thing a man spent so long building? Even if it hadn't been Remmy that built it?

Well when more people moved in, the housing authority

and Jim Johnstone hired him, Remmy, to build the houses next door. And since he could see this was going to happen, he bought up the lots himself and then sold them piece-by-piece to Jerry Holsapple, who then funded the building project. That way it got out of the hands of Jim Johnstone the Texarco company man, you see. And also he made a good chunk for the CAMELOT, MY MERRY MEN, AND OUR PRANK fund. Jerry Holsapple didn't really hate Jim Johnstone, cause Jerry lived not on the sharecropper side of the street but on the side of the street with the dukes and dames, so Jerry didn't know that Remmy had a reason for selling to him for cheaper than Jim would pay, he just said, "Okay, Remmy, if that's really what you want to do."

In building those houses, sometimes you'd have some dirt left over. There was a pile about fifty-five foot deep next to the well, for instance. They had reservoir water, but Remmy'd also dug a well, you see, and it was real good water. Inspector said it was good enough for a baby less than a year old to drink.

Well Jerry Holsapple's son Jeff and the Parish boy were six or eight years old. Older than Bren, for sure. "Can we play in your dirt?" they asked Remmy one day.

Remmy said, "Oh sure boys."

So the boys played in that fifty-five-foot-deep pile of dirt and worked it over and over.

They brought in more from one of the other houses, and those boys went to it again and worked it over and worked it over, their little boot prints in the dirt and some of their toys strewn about.

Round about that time, Remmy worked for a guy at a farm. And somewhere in the midst of that job, Remmy loaded up a dumptruck of sawdust that was full of cow manure – good for getting all those new lawns started. He brought it back to the big pile in the neighborhood, and the Holsapple boy and the Parish boy had brought along Jim Johnstone's son this time.

"Can we play in your dirt, Mr. Broganer?" the Johnstone son asked.

Remmy looked at him and looked at the Holsapple boy and the Parish boy, who were sharing a secretive grin. He didn't know the Parishes that well, but Jerry Holsapple would take it as a joke, of course. He'd laugh even though Mrs. Holsapple would be mad. But Jim would take it the worst way possible. A double joke then, one harmless and one for a target. So it was decided. "Sure boys! Go to her!"

And they worked that manure over and over and over, digging in it, playing their toys in it, getting it all deep in their fingernails and hair. Their mothers about killed Remmy, Holsapple and Parish both thought it was funny, but Jim Johnstone stared in through Remmy's front window, not even watching his own child as he hosed shit off his boy with water that wasn't soft.

WILSON REMUS

1965

HALFWAY HOME FROM Dallas (he'd given up on a vacation due to Oklahoma football traffic), he pulled over on the side of the road and watched the oil pumps for miles and miles and miles and wondered if he could have that if he just bought some up. It was tempting. All that money. Of course, he wasn't the only one with a ruined well. He wasn't the only one with a burnt shed. He wasn't the only one with holes in his walls where the wind could come in or with a toilet used by every stranger that went through the Texarco service station.

He stewed all the way home, thinking about it. And at last he said, "We're going camping instead." "We're doing what?" Beth said.

"Camping. It's a longer vacation."

The kids were excited, even though Mary mumbled about there being no State Fair at the campgrounds.

Beth looked worried. "Can you—"

"Course I can," Remmy said. "Course we can."

Beth ran the numbers the whole way there on a piece of paper.

She's the one kept the books, not Remmy. She still wasn't sure that they could.

Bren didn't care. He never had a merry men fund. He had his parents to fund everything he ever wanted. Money for him wasn't a renewable resource that came from working. Marionette was on the fence. She wanted to do it but she was worried about her dad, though she didn't tell them. So they packed up with the Holsapples and went out to the national park. Beth always bought lots of food, and they stocked up really good while they went out there. Three families out there — them, the Taylors, and the Holsapples. Pete Taylor'd moved away a bit north by then, you see, and they had to do friend vacations to make it work. It's not easy when your best friend lives hours away, you know. They had the green tents pitched and the little foil packs full of veggies and even the black iron rig for the peach cobbler they made big enough for a coven of witches, and seven cases of soda bottles. Well those two Holsapple boys grabbed ahold of those top poppers, you know, for soda and beer bottle caps. They were treating them like daggers for pirates, waving them around like fools. These were the long ones like you hang from a restaurant fridge.

Remmy heard something snapping and fizzing, snapping and fizzing, tinkling and tinkering. He got in there and looked.

Those two Holsapple boys had just finished popping all seven cases of soda. At once.

"Come here everybody! We gotta drink quick!"

They drank quick and burped a lot.

That morning, Mrs. Holsapple and Beth looked down at the ground and saw a fifty-dollar bill and they both shot down to grab it and bumped heads and started fighting over it, both of them had a hold of it so tight.

"It's mine!" Beth said.

"I saw it!" Mrs. Holsapple said.

"I saw it first!" Beth said.

"I grabbed it!"

Remmy shouted, "Ladies!"

They went quiet, both of them had ahold of it so tight.

Remmy said, "I have a clear and simple way of settling this." He walked up and gently took the fifty-dollar bill from them both. "Who saw it first?" he asked.

Both women said, "Me."

Remmy tore the fifty-dollar bill in half and gave each one half. "There you go, twenty-five dollars apiece."

Some of them laughed.

Beth said, "Dammit Remmy—"

"Is it a curse?" he asked.

"Yes it is," she said. "And I'll say it again. Dammit Remmy! Why you gotta go and destroy something we need just to get a laugh?"

"Well if it's good enough for King Solomon, it's good enough for me."

"Yeah, but this here's different," she said.

"I don't see how."

"At the end of the story, Solomon didn't cut that baby in two. He gave it to its mother."

Remmy stood stock still staring at his wife and what giggles remained subsided.

"Well mom," Marionette said, "that means he needs to give it to the tax man."

And the laughter started afresh, and Remmy loved his daughter at that moment even more than usual.

"You hush," Beth said, "this is between me and your father."

Remmy turned his back on his wife and walked towards the fire, winking at Marionette, whispering, "Good girl."

Marionette winked back.

Finally someone rewarding the effort he'd put into winking their way. Beth's half of that fifty-dollar bill they taped to the fridge and even to this day Remmy tells first-time guests that story.

Later, the girls all played euchre and Bren tried and got stomped and then they went to bed and the Holsapple boys went to bed and everybody went to bed but Pete Taylor and Remmy, who sat in those deep well folding chairs out under the stars and the dying light of the embers. You can see to the edge of the Milky Way out in them national parks if you sit quiet and wait out the last of the light pollution and watch.

Pete Taylor said, "You think we're alone out here in the universe?"

"Why, you feel lonely?" Remmy asked.

"Well sure. I don't have nobody up north."

They sat. A shooting star went by. Remmy thought of the meteorite. "Me neither, really."

"You got family."

"Sure."

"And employees and some friends," Pete Taylor said. "Holsapple seems nice."

"Sure. But I don't have you, Pete, and that's a lonely thing."

"Yeah."

They sat.

"Well do you think we're alone out here?" He pointed at the moon — how lost and dry and hard and cold and lifeless it seemed.

"I think part of living is being alone. That's how you know who you are. But I also don't think we're supposed to be alone. So you and me need to get together more."

"With the kids?"

"Or without them," Remmy said. "Either way."

"You still haven't answered my question, Remmy. You can't kettle pitcher a good question like you can a bad story."

"I think... what like aliens?"

"Sure," he said.

"Well of course they're out there, but it ain't like you think. Green men and all that."

"I just meant someone's as smart as us."

"Smart aleck more like," Remmy said.

The Good Lord snorted right in Remmy's ear and he jumped.

"Sure," Pete Taylor said to Remmy. He couldn't hear The Good Lord like Remmy could, not yet.

"I want to take that back. I didn't mean smart aleck," Remmy said. "But I did mean that we can't just up and think like this one day. You can't common sense your way out of common sense. That's why it's a gift, you know. Think of things like gravity or any old thing in space. Got there cause someone gave it leave to be there. Go back far enough and there ain't nothing that is that should be. All of it's saved like you rescued it from a burning building or a siege engine from castle and jester times."

"I never thought of it that way. That must be what my old lady thinks of when she says life's a gift."

"Sure. That's why you pray. Cause sometimes you want the cigar and sometimes you want the man behind the cigar that lets it keep on being what it is."

"Like home."

"Like coming to a place you never been before and finding it's home. Like the echo of the sweetest music you never heard. Like the scent of a flower you've never plucked. And every time I get the feeling and I go right back to those moments in my mind, I don't find the home or the song or the rose, but the memory. And when I remember them, then I get that feeling as if I started remembering a memory I've never come across before. As if I had a place in my mind for storing stuff that's true and connected everywhich way. Like a place just like a cave or an office in my mind filled with memories from some place I've never been." A bell rang out in his mind for a second time. "I long for it, Peter. I feel like a stranger in my own house."

"You and Beth okay?"

"What, like that? Fine. I mean… I can see this castle in my mind with the drawbridge up, and I want to get inside. And sometimes I do."

"Oh I've had that dream. Except for me it's a wine cellar."

Remmy laughed and threw him a Miller from out of the cooler. "It's like I don't want the castle so much as to become the castle. Or at least a part of it. I want the West Wind to fill the hole in my soul."

"Yeah," Pete Taylor said. "That sounds right to me, too."

"That's good to hear, Pete," The Good Lord said.

Pete lept out of his chair and wheeled around, having pulled his pistol and aimed it every which way. "You hear that?"

"Oh that's just The Good Lord talking," Remmy said. "You'll get used to him."

Remmy smoked his cigar. He never smoked except once in a blue moon, and he hadn't smoked in the last four blue moons and Pete Taylor was here and that was special so he figured it was okay.

Pete Taylor finally settled down and smoked his pipe. They made quite the cloud together.

"You think there's other people like us?"

"Well if The Good Lord didn't send that meteor then who did?"

The Good Lord said, "That's a good question, Remmy. And there ain't no reason I couldn't have someone send it just for you. Go deeper."

"That's a good point," Pete Taylor said.

Remmy said, "Dammit, Good Lord, just—"

"I ain't damning nothing yet," the Good Lord said. "Go deeper."

Pete Taylor cackled and sipped his beer and smoked his pipe. "Oh, sure, I could get used to this."

"Now that ain't fair with two of you," Remmy said. "It's bad enough Beth prays."

WILSON REMUS

1966

He couldn't get a rest, even with the travel. But he kept trying. He didn't have much money so he booked cabins around Lake of the Ozarks. He never packed his bag — Beth took care of that one. Away they went to the Lake. Just a cabin. Two burner stove like Beth was used to using. Running water and a bathroom, I think, can't quite remember for sure, but it was simple.

He went in to take a shower. Came out. He got into the bed in there. He was buckass naked as he was when he hopped up on old Rooney's truck.

Beth became aware. "Wilson Remus put on some undershorts."

But he was ramping up to making a point. "Don't have any."

"Well they're in the bag."

"Oh really?"

"Yes." She got out to show him. She moved the jeans.

She moved the shirts. She moved the stragglers from the Dopp kit. She moved every single thing every which way she could, but she could not conjure his skivvies. "I forgot'm."

"You did."

"Yes. I'm sorry, I'll—"

"You forgot my drawers."

"Yes, Remmy. I'm sorry, I'll—"

"My Goddamn drawers. Of all the things, woman, you forget a man's drawers. Why in the hell can't you think straight? Good God, what'll I do now, go commando like a whore? It's all your fault, ruining it like this."

The kids was staring at him shouting.

He got madder then. Not at them, but at himself for being so mad about such a pissant thing. Still he took it out on them by stomping outside and slamming that door.

He was quiet. And still buckass naked, just a kid out there in the elements and the bus had done left him behind again.

"Good Lord, work on her," Remmy said. "She needs help."

The Good Lord said nothing.

There he was again in that horse-drawn milk wagon, all of his toys before him and his father nowhere in sight and finding out they'd given Bloody Williamson to one another. He felt hopelessly lost and alone and feared for his soul and his bones.

"And I guess work on me too if I need to change," Remmy said.

"You talk to your bride like that again, Remmy, I will stop answering your prayers."

"Yes, Lord."

"Ain't got nothing to do with me not loving you. I love you Remmy."

"Yes, Lord."

"Ain't got nothing to do with forgiveness neither. I forgave you thousands of years ago, Remmy."

"Yes, Lord."

"Has to do with you being too stubborn to change your mind. Go on and change your mind now."

"Yes, Lord. I have."

"Good. Now go in and apologize before she's ruined for good."

"Yes, Lord."

"Remmy?"

"Yes Lord?"

"I'll always love you. But I still like you in spite of you acting like this."

Remmy cried.

"You tell some of the funniest jokes of anyone I've ever made, and your harmless pranks make the hard life livable. And you tell some of the best stories that make bad things into good things and there's nothing I love more than a good storyweaver. This is not about her and it's not about drawers. It's about you finding rest, and you will, Remmy. You're getting there."

"I feel so lost," Remmy said. "If you just gave me a bearing or a map or a compass or maybe one of them triple-A trip guide thingies we used on the way to Dallas that one time. Then I'd know the way to the place you're sending me."

"I'm The Way," The Good Lord said.

Remmy cried again.

He went back inside.

The kids was bracing themselves.

Beth wasn't crying at the moment, but her mascara'd been running.

"I am an asshole," he said without preamble.

"No you're not," she said.

"Well then I *was* an asshole and *acted* like an asshole to you."

"Okay," she said.

"Sorry," he said.

"Okay," she said.

"Forgive me?" he asked.

She nodded.

"Forgive me?" he asked the kids.

"Yes daddy," Marionette said. She fumbled with the nail polish she'd been using so that she accidentally painted the whole tip of her finger before she'd really noticed.

Bren ran up and hugged his leg. Any taller, the kid wouldn't be able to do that anymore. The kid probably got the tall genes he hadn't gotten. Probably got the long jeans as well.

He went commando that night and bought some shorts later and it was fine. He opened a can of acetone for Marionette to dip her finger in, and she started over. He even got some good jokes about it when the girls weren't listening but Bren was. Jokes about long genes.

For some reason, not many people bothered to water ski on that trip. He didn't get it. But Remmy loved to ski so he went in the boat with everybody and he went round and around on them skis. When that old boy cut the power, the boat had hit on some rocks and needed pulled out by hand. Remmy sank down to his skis and he was on rocks too. About a foot deep. He was almost walking on water, but there was rocks just a foot down. Had he fallen on his skis, had he looked down for just a moment, he would have been beaten and skinned alive.

He helped pull out the boat and gave up water skiing for a while.

When he got home, Jim Johnstone was working in his garage on his car, his back to Remmy, the hood open. Old Texarco collectibles and signs hung all around that garage, as you'd expect from a fully devoted company man. And Remmy squinted and noticed some tools in there that looked hauntingly familiar. Like they'd belonged to him at one point.

WILSON REMUS

1967

NEW JOBS GOT bigger for Remmy. Bigger and bigger until he really did start to cover the map with the specks that were his houses. His and Daddy John's. But mostly his. Half of his, mine, and my brother's friends' houses growing up were built by one or the other of them. Daddy John insisted on giving him the credit since he hadn't wanted to take the risk in the first place, but in Remmy's mind they'd built them together with all them other boys, fair and square.

Sometimes he'd get up at two in the morning: get up, couldn't sleep, went back to the office, and worked on things. The worry kept him working. So'd the anger. Sometimes he found himself thinking about Jim Johnstone and he'd get so pissed off he couldn't get to sleep, and so he'd go to work again at no soul's hour. It was a bad habit to get into, but it was about the only vice he had.

He still needed a break, so they tried going out to

Cayman Rock on the Ohio River. They went down there ten times in two years, and when they started fishing, Good Lord, Beth really got out there and fished. Man she got fish. She pulled them in just like she was doing nothing but shopping at the grocer or ringing a dinner bell. Bren caught a big old sauger probably the size of a bird dog, caught it using Remmy's method of feeding it your own flesh and blood, and Beth tried to grab it and he refused to let her kill it. He told her he was taking it home to be a pet.

"Pet? What you need a pet for, Bren?" Beth asked.

"Them's good eating," Remmy said. "Cut you some flank."

"No sir," Bren said. "I am a game fisherman."

They laughed at him.

Marionette laughed late. Hers was just for spite.

"I am. I just decided," Bren said. "I am a game fisherman and I will only hang onto them long enough to put in Carlyle Lake by our house. You watch. One day we'll have a big old sauger population there at Carlyle Lake and downstream in the Kaskaskia." He lifted up that big old fish and stumbled a bit under the weight. "See her? She's pregnant and those babies will be good food for the other fishes and some of them'll make it and get big and be good catching for when I want to feed them my own flesh and blood later."

"Alright! Alright!" Remmy said. "I retreat! We'll take her home."

"Good," said Bren. And he plopped the fish in the empty cooler and filled it cupful by cupful with lake water. He glanced up.

Beth had a whole slab of sauger and bass and catfish stacked next to her, and she was slicing off large hunks of meat and descaling slabs of skin and emptying guts and fishbones and fish heads into a great big pile next to them. And Marionette's hauling them to the water as scraps for the bottom feeders.

One of the grafts of fish skin and scale landed on Bren's face as if he'd been a netherworld dragon born with the caul. And he screamed.

They'd come home and Jim Johnstone'd be working on his car in a garage filled with Texarco signs and Remmy's tools and water cleaner than Remmy's well and sleep in a bed paid for by the people like Daddy John that had sold out their land cause they couldn't afford no other and under a roof built by his contractor brothers, and Remmy's vacation would wear off before he pulled up the driveway. Jim Johnstone didn't ever admit it. He just bragged about how he got heirloom seeds for his garden out of Rickers and how he was growing all sorts of things. Not just vegetables. Poppies for opium. Marijuana. Bloodroot and hemlock.

He said *hemlock* like it was a threat or something.

WILSON REMUS

1968

MARIONETTE WAS TWELVE. She was dating a big tall boy. Six-foot-two at only fourteen — he'd end up being almost seven by the time he was twenty-one. And she was barely four feet. That giant of a boy's gonna have an awful time trying to kiss her, you know? Marionette tried so hard to impress the boys. Remmy didn't see why: the daddy's the only man she need impress in her life, right? He wasn't around much on account of the working toward the rest, but — well shit. He realized when his daughter brought her boyfriend home that he was reliving Daddy John's life.

That wouldn't do. He needed to find his haven and quick.

Meanwhile, he'd make it up to her. He rigged it all up nice on the porch there in the dark, waiting for that boy to drive his Ford into the driveway and walk her to the front where he'd kiss her just like Remmy used to do when he was dating Beth. Well, not Beth exactly. More like the other girls before Beth and he woulda done Beth that way too, but Beth's hard to get thing got in the way, though it made her fun for the catching.

But Marionette and that giant boy, well of course they'd want to be kissing. So Remmy waited in the dark and heard them talking coming up the walkway, and he waited until they were right up on him and threw the spotlight on the step ladder.

"Dad!" Marionette said.

"It's okay, Mary, it's okay," Remmy said. "I made you a ladder so he can kiss you."

"DAD!"

"Just step up here."

Oh that giant boy was giggling.

Marionette got mad and commenced to punching him. The giant boy, that is, which didn't make a whole lot of sense. Well that boy picked her up and walked her over to the step ladder and set her back down there in the limelight Remmy'd rigged, and damned if they weren't eye-to-eye.

"See?" Giant Boy said. And he kissed her.

And she punched him in the chest.

"Oh come on now," Remmy said, holding the light. "He looks like a good kisser, Mary."

She muffled another *Dad!* between the lips, but then she yielded and she went to kissing him back. Kissing the boy hard. And the boy tried to pull away but she wasn't having none of it, those men trying to manipulate her like that, and it got to where he had to decide between kissing her and yanking her clean off that ladder.

So he kissed her back and it got to be romantic.

Remmy smiled.

Then it got to be a little *heated.*

And Remmy said, "Now now," and turned off that spotlight.

"Your idea Dad," she said.

"All I said tonight was that I'd help you get kissed, Mary. I didn't say I'd help you start a family at twelve."

That giant boy was a gentle giant, but after that night, he stopped showing up. Remmy was afraid he'd scared him off. Or she had.

He found himself missing his friends who he'd lived with out by the oil fields before the derrick ruined and contaminated and adulterated and befouled everything. He recalled the way Grandad Patrick had fought those companies, even fighting them through a loss until he worked them over good or at least got the last say of it. He recalled the pain of losing Toby to pneumonia and the joy of the kids and the whole world seemed to open up to him and show itself hanging there on the coat hooks of God's master closet.

He took a left by a fallen tree. Cloud and canopy cleared and the sunshine glowed and gleamed. A glade there had hid the minarets of a mansion. Maybe a castle? Some old pioneers had built a fort out here for wars Civil and uncivil, using it like the old Alamo or the old garrison at West Point. He saw the eye slits in the face of the tower where the arrows could go. He saw the walls where his family and men could hide.

And the Good Lord said, "Enough of a sign for you?"

And Remmy said, "Thank you, yes."

He climbed up there and started looking around.

Dwellings which could be fortified into right and proper houses. Places for horses or cars. Places for eating large family meals. Places for the kids to play, and they could make themselves a little camping paradise out here, now couldn't they?

He did the figuring. He could sell this next house and have enough saved to do it with. He could buy it up and give it to his buddies at cost if they'd just come and move.

They'd come. Sure they would. He'd found Camelot, hadn't he? Now for the Merry Men.

And then their great big impune prank.

WILSON REMUS

1969

THEY HAD COMPANY over and Beth made up some French dish Remmy couldn't pronounce, but it tasted like fancy beef stew. They all sat down together, and it was the Holsapples and the Parishes and the Johnstones. Beth was trying to keep the peace, you see. Getting them to eat together, you see.

Remmy ate about half of his when Jim got to running his mouth, which was not a good start for keeping the peace.

"You buying any more land, boys?" Jim asked.

"Some," Jerry said.

"You, Remmy?" Jim asked.

"I might," Remmy said. "I've been thinking about it." He still hadn't gotten commitments from his Merry Men for their Camelot.

"Whereabouts?" Jim asked.

Remmy looked at Jerry.

Jerry raised his eyebrow.

Remmy said, "Not sure yet."

"Well I just signed the papers this morning on the most wonderful tract of land. It might have oil value. Might have lakeside value, if they expand that cove. Might have subdivision value if it comes that way."

Remmy suddenly listened real close.

"Oh yeah, boys. Nice old forest plot. Old too. You know what I mean by old? Has history to it."

Remmy'd set down his fork and stared at the man.

"Right in the middle — this is what sold me on it. I didn't know it was there until I went exploring the other day — but right in the middle is this old fort. This beautiful old castle, isn't it Maddie?"

Maddie said, "Oh yes, Beth, you have got to see this thing. Like right out of a story book."

"Is it?" Beth asked. "I reckon I'd love to see it."

Remmy hadn't told Beth about it, you see. He wanted to spring it on her by surprise.

"Shame it's in such disrepair though, you know Beth?"

"Oh sure," Beth said. "You have to tear down so much just to get started."

"What are you fools talking on about?" Remmy said.

The table got really quiet.

"Well Remmy, we's just talking about an old lot and an old bunch of stones," Jim said.

"Acting like we just throw it all away. If a house has good studs, that's the best thing a man can hope for. You can fix any house with good studs."

"You'd have me fix up this castle?" Jim said. "That's

an expensive task, and there's not much call for castles any more."

"Well fine then," Remmy said, "you don't want it, then sell it to me."

"You'll pay twice what I paid," Jim said.

Jerry cleared his throat and coughed.

Remmy eyed him.

"I'm sorry, Remmy," Jim said, "but a man's gotta make a profit."

"I thought you made plenty on oil," Remmy said.

"Well yes," Jim said, "but the oil don't factor in here, you see. I bought this land and it would just be irresponsible of me to sell it for less. Aren't I a steward of the Good Lord's money and land?"

The Good Lord said to Remmy, "He acts more like a brigand duke than a steward of The King."

Remmy nodded.

Jim thought he was nodding at him so he said, "You see, then? Double. What do you say?"

"I say you're full of alfalfa," Remmy said.

"What's that supposed to mean?" Jim said. "I just want you to think about it is all."

Remmy thought about it. No way he could afford it, not with even all the money the bank'd loan him. Hell, the first price was the outer limits of what he could afford. "I'll think about it."

"Well think quickly cause I hired those boys to tear it down tomorrow."

"You what?" Remmy shoved his plate forward towards the center of the table.

"Now boys," Maddie said. "Ain't no reason to talk business at a party is it?"

"That's right," Beth said. "Maddie? You help me with these dishes?"

She did.

They went in there.

"You act like it already belonged to you," Jim said. "Like it's yours."

"For a man that uses the word steward, you sure screw up freehold and leasehold."

"Whatever does that mean, Remmy?"

Remmy said, "I mean you think you own whatever you touch. I'm here to tell you I don't own nothing."

"I'm sorry to hear you're in bad with the bank, but that ain't—"

"You listen here and listen good, Jim Johnstone. I paid off my house. Both of them. First one to you and your company, now this one. I ain't in debt, but I also don't think it's my house. Ain't none of it mine. All of it's a gift to share from the nicest castle to the pile of shit your boy played in."

"I knew you were in on that," Jim said.

"Told your wife I was."

"I see," Jim said. "I'll have to talk to her about that."

"You do that," Remmy said.

Jerry was sipping his soup really loud and they both looked at him. "Look boys, I think the football team's doing better."

It was quiet for awhile.

They made it, somehow, someway, into the family room — that nice new one Remmy'd built inside of which he now knew Jim Johnstone'd been several times as a thief.

They talked about grass.

The ladies talked about fishing. And then they talked about some man that was dating a black woman.

"She was African American," Remmy said from the other room.

"No, she was a negro," Beth said.

"That's the same thing," Remmy said. "And mine's the kind way."

"No it's not. She was negro."

"I can't win with you," he said.

They went back to their separate talks.

The boys talked about earth movers and something called Woodstock. The ladies talked about how much and what they'd been reading. Jim's wife talked about how she'd only buy a certain brand of handbags in just the snobbiest whiny voice you ever did hear. She talked on and on about hanging out with the other wives of Texarco chiefs.

So Remmy told a poop joke with her as the subject.

That didn't go over well, but it made sense to Remmy. Truth is that all humor is scatological at heart. The word "humor" comes from the same word that gives us "humidity." As in moisture. Bodily fluids. That's what Southern Illinois boys mean when they make fun of snobs and divas and say, "Your shit stinks as bad as everybody else's." There's this sense that, if you can't laugh at yourself, you're screwed, cause your shit's just as funny as everyone else's shit.

But Jim Johnstone didn't get the joke and neither did his wife and there was a fairly lengthy awkward silence. That's the problem with a jester's targets: sometimes they just don't get it.

Remmy just shook his head and laughed anyways at them all.

Somehow, someway the talk resumed and got back to full steam.

Remmy had to get up at 5am the next morning. He wasn't a banker or a land owner or an oil man. He was a carpenter, and there ain't nothing worse in carpentry than the heat of the noonday sunshine while you're working in hot roofing tar. They stayed and they stayed cause they had to go to work at 8am. They stayed. Remmy had to get up at 5am. They talked then about picking plums and stayed. Remmy said to himself he didn't care and he got up and they stared at him.

"Yes Remmy?" Jim said.

"Goodnight." And he walked out of there and went to bed.

Sometime much later, he woke up to the rustling of Beth crawling into the far side of the bed, and he reached over to touch her, but she was cold and stiff and unresponsive to him. He sighed and rolled over.

Next day, he got an earful from her, and he didn't even bother excusing it. She invited that whole crew back over the next week to try again, and they all came over, and Remmy suffered through the small talk of that old fool and told old stories about when he and Jerry went to school together for a brief time. They went into the family room and he got tired and instead of getting mad he said *I'll show them* and fell asleep right in his chair like Daddy John and Grandad Patrick would do him.

Man it felt good.

In the morning, he woke to find a tent in their front

yard. And a fire pit with a fire and a hog on a spit slow roasting over it. Dirty clothes hung out to dry on a clothesline in front of his favorite oak. Whittled wooden pikes with bloody war clothes on them. Or maybe painting clothes with red paint. And that massive tent with a flag on it, like the standard of some dark knight.

This wasn't the work of no devil like Jim Johnstone, company man. He didn't have the imagination.

This was the work of a trickster like me and him.

He went right back into the kitchen and said, "Beth?"

She was mixing muffins and didn't look up. "Mmm?"

"Beth? Beth?" he asked.

She looked up. "What? What?" She gave him her frown.

"Pete Taylor's in town."

"How do you know?"

He pointed out the open door.

She leaned round the corner and saw the base camp of the rebel army strung up in her front yard. "Oh good grief, Peter."

"I'll get even with him for that."

He went to work and did his hammering on some shingles, laying them one at a time all alone with his help elsewhere, and on the way home he saw a shitty old small engine block sitting out in the yard with a SCRAP FOR SALE sign on it along with some others, and he got curious. He went up to the front door and asked if they had a big scale where they could weigh the engine block, and the guy said he already had, since scrap was by the pound.

"How much it weigh?" Remmy said.

"Exactly three hundred and thirty-three pounds."

"Perfect," Remmy said, "I'll take it."

Well Pete Taylor'd gone in to their house while Remmy'd been out working and buying scrap, and when Remmy got home, his front door was gone.

"Pete Taylor's in town!" he shouted again through the open doorframe. That made two on him. So Remmy started looking around at the neighborhood, standing there in the doorway scanning for places that would hide a Pete Taylor. He looked over at Holsapple's house. Holsapple would do it. Remmy knew that house well cause he'd sketched it and built it himself. He went over there with an M80 because just sure as shooting skeet that Pete Taylor's hiding on the other side of that fence, hanging onto the Broganer's front door. So he lit the fuse of that little Fourth of July surprise and tossed it over that fence and KA-BOOUUGHK. Dirt, it shot up. Paint off the siding, it shot up too.

"Good God, Remmy!" Pete shouted from the other side and came out limping.

"Why Pete," Remmy said from the safety of his front doorway, "I didn't know you was in town. Something wrong?"

"Don't give me that, I know it was you. You're gonna blow my toes off!"

"Why you limping?"

"I did that the other day."

"Bullshit. You rolled your ankle dodging my M80 there in the backyard, so let's stop lying to each other and where's my door?"

Pete Taylor hung the door back and everything. Remmy needed a sneaky man like Pete on his team.

"You boys quit it with those bombs," Beth said. "Someone's liable to blow off half a leg and bleed out on the street or lose an ear."

"That's absurd," Remmy said. "This ain't no air raid."

"And it ain't no ear raid neither," Pete Taylor said. "We're much more likely to blow off a finger or a foot."

"Oh that's much better," Beth said. "Try swinging a hammer then."

"Only need one swing to strike the brightest bell," Remmy said.

"But if you lose your middle finger, how you gonna flip people off?" Beth said. "Cursing because of a curse my foot. Couldn't even say that if you blew your foot off, listen to you two. Get inside."

He tried and tried to get Pete Taylor to move back to town — to move into the neighborhood — but Pete never budged.

WILSON REMUS

1970

THE IDEA CAME to him that he could invite all kinds of friends to build his Camelot right there in the neighborhood. He didn't need that old fort out in the woods, the one Jim Johnstone'd stole from him. He could just try to persuade them to move into *his* neck of the woods. He went hustling like crazy, he did. He asked Hellman and Rooney and Norm and half the town. He asked his brother-in-law and got him and Gwen to come move next door. Every time someone moved or a lot opened up, he'd buy it and sell it for way cheaper than Jim Johnstone'd sell it. Good chunk of people told him 'no,' but a good chunk did and before long he'd loaded up the sharecropper side of the street with all of his favorite carpenters and oil field workers and mechanics and what have you. He had his Merry Men.

That summer, he used a Louisville Slugger on a man. Guy named Sinclair and his son were working for him and Daddy John, doing some of the electric work. Well

Sinclair's son came on a big old wire that was hot and he got to shaking and popping there. The clicking sounded like a grasshopper the size of the courthouse had come to eat their studs.

Remmy recognized the signs from Beth and the stove Jim Johnstone's company'd sold him. So he took the baseball bat that he'd taken out of the trunk, the same one he'd smashed apples with in Pickneyville, the one leaning on the wall, and pushed at the boy. He knew it wouldn't conduct and had experience with breaking contact on account of the stove.

Well he pushed at the boy and pushed at the boy and the boy wasn't coming loose so he reared back and swung for the fence right on the biggest muscle he could find, which was the boy's back. The kid blasted forward and turned loose.

"Sorry, Sinclair," he said to the kid's father. "I didn't know how to get him—"

"Thank you. Oh God, thank you," Sinclair said.

The boy took the day off and came back the next day to finish the work, bowed double, which left Remmy impressed. A few days later on another house, they were building a basement, putting footing around it to be able to fit a form for the walls. They kept begging to get out of that basement. Hot as hell.

The telephone rang. Remmy had a telephone on his desk and he always kept a phone tied into any job site. He wouldn't do the war trailer thing like Jim Johnstone done — he thought that was heinous. But he tied in a phone and the phone rang then.

"We don't need to finish it," one of the boys said on the other end, the worksite end.

"We need our finisher to strike it off with a two-by-four," Remmy said into the phone from the comfort of his desk.

"Don't finish," said Sinclair.

Remmy's brother-in-law Ryan had come down there to the site that day and brought a thermometer. He'd been at this house the day before, hot as it was he wanted to know how hot it got down in that basement. He held up that thermometer then took over the phone and said, "Well hear this, Remmy, it's a hundred-and-fifteen degrees down here."

A basement half-built in the summer heat doesn't work like a cool wine cellar. Works more like Death Valley.

The cement finisher picked up the phone and said, "I will not work in this inhuman conditions."

"If you can spell *inhumane conditions,* I'll buy you one."

Then the cement finisher got the rest of them to shout, "WE WANT BEER! WE WANT BEER!"

"I'll buy you one as soon as you're done," Remmy said.

So they went on strike. All of them. They dropped their tools.

"Oh now come on," Remmy said. "This ain't Bloody Williamson."

"You're professing us!" they shouted.

"Oppressing. Really? You're really gonna tell me that I'm working you to the bone like one of them oil and coal companies? Good grief, fellas, that's low."

Sinclair took the phone and said, "Then buy us a case of beer and give it to us."

"Let me think about it," he said, and hung up.

The phone rang.

It was the concrete truck. He had more concrete coming. He hung up.

The phone rang.

"We will not work in this," his brother-in-law said.

"Dammit, Ryan."

Sinclair took the phone again, "Buy a case of beer and give it to us. It's the least you can do for electrocuting my boy."

"Now wait a minute," Remmy said. "That's too far."

"You're right," Sinclair said, "that was mean but have a little mercy, Remmy."

"You still think I'm as bad as Bloody Williamson?"

"Well of course not. But a man needs a beer in heat like this."

Remmy said, "Well okay. Don't die now."

Cheers sounded on the other end, and he went and got a case of beer and took it to the work site. Sinclair's son, the one got electrocuted? He was the worst one. He guzzled it down, twelve bottles straight, and passed out. Remmy and Ryan and the rest of the boys had to carry him out of the hole and set him up against the tree.

"Don't you ever bring another thermometer to work," Remmy said to Ryan.

Ryan raised his beer.

They struck it off, and the cement finisher'd left and so the boys had to finish it while the Sinclair boy slept against that old oak. While finishing, Remmy said to Sinclair, "Don't ever send him out on another job like this."

Sinclair had two sons. One died as a twenty-year-old.

The one that died was a cement finisher. He said then, "Day like this, it's safe to say the wrong son died."

With his dream castle stolen by Johnstone, Remmy had to do something that would bring him closer to the peace and rest he longed for – his Camelot. Maybe it would help him find the world's biggest impune prank up there in the sky.

It didn't do either. But hell, it was fun to fly.

One time, he got a little too drunk and hopped in his little prop plane at the airport and woke up in Chicago. He said, "Good Lord, I'll never drink and fly again."

"And what about drinking and driving, Remmy?"

"That's a good idea. Probably shouldn't do that either."

No, at fifty thousand deaths a year in 1970, he probably shouldn't do that either.

Pete Taylor called one day, working out of town for Texarco at the time, along with Ms. Taylor. So he showed up at the doorstoop, and Marionette told him Remmy was over at Stan's. He called there and said, "What are you doing?"

Remmy said, "We're gonna play cards pretty soon, and we're gonna get tacos."

"You're having tacos?"

"We made 'em ourselves."

"Did you now?"

"Well yes, Pete, you know I'm a fine taco chef."

"Send me one!" Pete Taylor said.

So Remmy took one, shoved it into one of Stan's wife's manila envelopes, put a stamp on it, and mailed it. Got to Pete's house weeks later, and Pete had that old taco framed and put up on the wall like a rookie baseball card.

The girls decided then it was time for them to figure out how to see each other more often.

So. Another use for an airplane.

The flying club was him and Pete Taylor and some of the other boys. They started out renting a plane together, paying on it together, a bigger one they could take places. They waited until they'd all saved up enough to buy their own planes and bought all at the same time. See, if one of them wouldn't have waited, the others couldn't afford to keep paying on the big plane less one share, and they'd all have had to quit, and then the worst thing's flying alone forever with no one in the canoe in front of you.

One time, Pete Taylor said, "Remmy, you think we can outrun eagles?"

"I think I could outrun peregrine falcons," Remmy said.

"What do you say we try?"

"How so?"

"I got a idea."

Remmy had one of those prop planes that was a two seater or so. Pete Taylor couldn't find no falcons. Couldn't find no eagles. Couldn't find no red-tailed hawks or owls. Couldn't even find a flock of Canadian Geese, which is really unlikely in Southern Illinois during the winter, since they all migrate down and eat everything and shit everywhere.

Anyways, Pete Taylor got ahold of seven chickens and had a tight coup he brought up with them into the air — you know, a box of chickens on revolt. The idea was to toss them out the plane, let them fly for awhile, and then chase after them to see if they could catch them. This was dumb. Dumb, dumb. For one, other than the little chat about Orville Wright's bird strike and Howard Hughes' poultriless crash back in 1946, Remmy'd never stopped to consider that feathers don't mix with engines, when matching wings to wings.

Petey dropped them birds out the side of the plane and shouted, "GO TO 'EM REMMY!"

And Remmy pushed on that stick and dove and then saw the way they was barely flapping and spiraling and remembered something his daddy'd taught him once on that beautiful farm so long ago.

Daddy John had said, "Chickens cain't fly for shit." And that was the other reason it was dumb.

All seven of them birds freefalled straight down and landed *splat* in seven red bangs on the parking lot of the factory, luckily on lunch break.

"Whoops," said Pete Taylor.

"Forgot," Remmy said.

And they got the giggles and could barely keep the plane straight enough to land.

Remmy stretched out his arms in later years and said, "There's nothing like flying." His eyes were in a different place whenever he said it. "There's nothing like being up in the sky and feeling the wind play with the stick and knowing you *could go anywhere*. And yet… I never got quite high enough."

His castle in the clouds, that plane. Just like *Jonathan Livingston Seagull*, which he read that year. He took Bren and Mary and he'd do corkscrews. He'd take Beth and she'd punch him in the shoulder when he'd kill the engine and just let them fall — punch him and punch him until he'd start her back up and cackle like an old trickster fool.

WILSON REMUS

1971

REMMY HAD TO join the Junior Chamber of Commerce summit up in Chicago. At first, Pete Taylor and Remmy decided they was just the two of them gonna go. But then they thought about all the fun they could share, pranking and making a ruckus and decided they wanted the guys in Bellhammer and Salem and Carlyle all to go up together. So they rounded up the sharecropper side of the street and some of the other boys.

Oh they dressed up something fierce.

Tan shoes. Pink shoelaces. Polka dotted vest and a purple hat band.

Man oh man.

Anyhow, they dressed like what they thought that song was saying, you know the song that lists all that stuff. Dodie Stevens sang it. So they went to the flower shop and got a purple piece of ribbon and put it on their hats, and Beth made some vests and they put pink shoelaces with some tan shoes and wore it all on the train to Chicago.

They shouted about it like a bunch of hooligans.

Everyone wanted to give them some cigarettes to advertise for them, something to put in their vest. Local rollers. And county rollers. And then state rollers asked. And Phillip Morris called when they heard about it and asked Remmy to put some of their cigarettes in his pocket, and Remmy said, "Your cigarettes are made out of horse shit and alfalfa."

And Pete Taylor said, "There's not a damn bit of alfalfa in them."

They went with some local brand and asked them to get paid to advertise, and the rollers said yes and paid them all enough in cash that they pretty much got their meals for free.

Then at O'Fallon, twelve other guys got on with tan shoes, pink shoelaces, polka dotted vests, and purple hat bands. They all laughed and went together — a mob right out of a vaudeville act — up to Chicago for the Junior Chamber of Commerce summit. And Pete Taylor and the boys played all kinds of tricks on that other crew.

Well they got up there and settled in at the YMCA, Remmy and his Merry Men, and across from them was that barber's boy, who didn't look very good and especially not that he would someday amount to a barber himself. That kid was out and about going into those convention rooms and asking them all to vote for him. He hoped to work that whole crowd, but somewhere in there Pete Taylor found out the kid wasn't much of a drinker.

Well they bought screwdrivers, sipped them, set them down on the piano near where the kid was talking, and walked away. Remmy saw the kid seeing them there, and since everyone else had a drink in his hand he grabbed one thinking it was orange juice or something and then walked around with it in his hand, sipping it from time to time. When it was gone, he put one down, came back to the piano, had another, and so on.

Remmy came up to him halfway through and said, "Hey son, how's the crusade going?"

"Well Texarco sponsored me, but I'm not supposed to say that. Gave a bunch of money to the Army Corp of Engineers again, but I'm not supposed to say that either." He sipped.

Remmy was mad, then. Not at the boy, but at Texarco. He'd show that company what he thought of their candidate. He bought a Bloody Mary. Pete Taylor put extra, extra vodka in it. They gave it to the kid at the next place and told him, "Normally they just drink these in the morning."

"Is it very strong?" the kid asked. He was pretty deep. "I can't have much more."

Remmy said, "Try it and see if it suits you."

The kid tried it. It suited him fine.

They kept walking the town and getting him drunker than a skunk. They got him back to that YMCA finally, careening down that hall like the old milk wagon when it'd lost a wheel the year before Remmy'd left and Daddy John had shouted, both of them like Bloody Williamson. And when that boy laid down in the hallway, he went right to sleep. So they got a big old king-sized sheet, they rolled him up like a big old burrito, and Remmy carried one end and Pete Taylor carried the other.

They took him six rooms down and let him lay there all night.

Big shot wanted to be Texarco's public servant. So they just sent him back to barbering, where he belonged.

WILSON REMUS

1972

REMMY WORKED NIGHT and day and trying to buy up and build up his neighborhood with the boys. They got away for a weekend sometimes. They went back to Canyon Rock, held those fish fries at the Ohio River. Just an ordinary picnic. They rode the ferry across for a nickel. They went to Washington, DC that year too, when Bren was 12. They stopped at every wide spot in the road to pick up beer cans to cash in at the recycling place for extra money on the trip. Bren and Remmy went looking for beercans inside the fence of this seedy hotel. Ten foot fence at the hotel, and they had to buzz in. Over in the corner of the fence, Bren saw a cache of fireworks laying up on its side.

"Free fireworks!" he shouts.

Beth looked at Remmy.

Remmy shrugged.

Marionette whined about going home.

Halfway through Bren filling up his backpack with bombs and Remmy holding the rest of the box, guy comes running up screaming. He ran a stand down the road. Seems somebody stole the box and then dropped it, right where Remmy picked it up.

The guy called the cops.

Bren was sure *he* was gonna get arrested.

Remmy talked them out of it. Talked just like he'd talked back when he was racing the Magic Chevy. They gave the fireworks back, and it was all okay, but Marionette wouldn't stop whining the rest of the night. She whined and she whined and she whined so much that Remmy said, "Girl if you don't shut up I'm going to buy you a bus ticket and send your ass home."

"Good," Marionette said. She wanted to be with her boyfriend, you see.

But Remmy's not the kind of poker player you go all in against unless you know for sure what hand he's holding, cause he just might have you beat. "Alright then, Greyhound it is."

"Oh good!" Marionette said.

Now it was Beth's turn to complain and call Remmy crazy. But Remmy wouldn't be dissuaded.

He got to the bus station, and since Marionette was fifteen and didn't know any better, he bought her a ticket to New York City and tipped both the ticket master and the driver to keep an eye on her. Then he bought a ticket from New York back to Indianapolis so that it'd time out that she'd arrive just when they finished with their trip. "Get her back safe and I'll tip you again," he said. He turned to Marionette, who was plain ignorant of what kind of ride she was about to take and said, "All right honey, we'll see you at home."

"Oh thank you, Daddy!"

"Sure," he said, grinning like a beaver who'd just built something massive.

She wasn't so thankful when she finally got home after sitting thirty-one hours in a Greyhound Bus seat, but she never whined on vacation no more.

Pete got president of the Junior Chamber of Commerce that year again. They said anyone that's that old being president of a young men's club has to be a dead chicken. So they gave him fifty dead chickens and made him fry all of them for them. They called him Dead Chicken after that.

They went to Chicago for a JC convention — just the two of them, this time. Whole bunch of JC's there from all over the state. They went into this hotel that would be torn down in 30 days.

One of the guys from Mt. Vernon said, "Look at all the keys there for the rooms." There they were, just hanging on the wall behind the counter, and that group of boys looked at one another and then Pete Taylor hopped up on the counter and passed out 400 keys to 400 rooms to Remmy and the rest of them. And everybody started going around to empty rooms, and in six rooms people were still in there having sex or whatever, but the rest of the rooms they got stuff out of.

In particular, they got all the pillows.

800 pillows from 400 rooms. Maybe more from the doubles. It was a metric shitton of pillows.

Well, they went up to the top floor and started ripping

the feathers out of all of them pillows — there had to be twenty or thirty guys on the roof drunk as skunks and crazy as cocks. They ripped up them goose feather pillows and dumped them out into the air and onto the street, and then they ran like crazy and watched as the wind made it rain down angel feathers on the black heart of Chicago. They got two blocks away and two old boys with pushbrooms came out to start cleaning it up.

Remmy spotted a bunch of longer feathers like you use on fletchings — just a bunch of them had no business being inside featherdown pillows in the first place, and he grabbed a whole sackful and carted them all the way home.

That year the Taylors and the Broganers went down to Florida for a vacation without the kids. They didn't do it all that often, but they did it that year. Gulf Coast. Nice weather and blackened grou- per. There was a state park, new one, used to be a zoo that held the world's largest hippo. Well when the zoo became a park, the people protested cause they wanted to keep the hippo, the crown jewel of that area. Biggest crown jewel any town ever had.

Well they got what they wanted, and the hippo stayed, and the tours resumed, and that all happened about two months before Remmy and Pete and Pete's wife and Beth all showed up in Florida. Beth was wearing her best white blouse, white capris, white hat, white handbag, all very Florida. They heard about the hippo from some of their RV camping neighbors who talked it up, and all four of them hopped in the truck that Pete had drove and went off to the park.

Well there was a large crowd on that bright and sunny day there at the hippo park, and the tour guide was standing up in the enclosure of this monster talking about how many metric tons of cabbage it eats, how many swimming pools full of water it drinks per day. You know how tour guides can drone on about trivial nonsense you'll never remember and don't affect the price of sorghum molasses.

About this time, Beth had gotten into taking pictures. She deemed herself a regular old photo documentarian. Never mind how her finger blocked the flash half the time and the lens the other half, she was a master. She got a visual record of everything. Albums and albums and albums of photos. Gosh they probably spent a third child's birthright — Toby's birthright — on developing rolls of film and whatnot.

So, she's in the middle of that crowd with Pete's wife, bound and determined to get a picture of this slack jawed monster that eats metric tons of cabbage and drank swimming pools of water. If you've ever been in a big crowd at a mosh pit or if you've ever been pushed around while some public servant shakes hands and kisses babies at a rally, you know that crowds work like tides and cur- rents, just like living water. And if that tide and that current pushes you, you stand about as much chance of stopping it as old Cnut. The current of the crowd took her and Pete's wife to the back end of the hippo.

And hippos do this thing where they *plrlrlrlrlrldldlddt* with their tail over their asshole and fertilize the land with their shit.

Soon as that guy starts talking about how many cabbages it eats, old Big Berta the hippo reared up and just painted Beth and Pete's wife with hippo shit. Covered Beth especially, all that white turned diarrhea brown.

And Beth gave the most normal response imaginable which was: "FLEAAHAAAAGH!"

And Remmy said, "GET AWAY GET AWAY GET AWAY BETH!" And he and Pete Taylor sprinted to the truck, hopped in.

And locked the doors.

WILSON REMUS

1973

THEY GOT SNOWED in, and Bren was desperate to go hunting with his new bb gun, so Beth took a bunch of old pillows and placed them at the end of the hall and put a box in front of it and painted an oil derrick on it and Bren went to shooting in his own private shooting range.

Even though the kids were older, Remmy thought he'd take them to see Disney's *Robin Hood* — the one with all the animals. They had a family meal one night and then went out for popcorn like Grandad Patrick used to do for him back when Beth worked the theater and he sat down and watched it with them. And damned if that didn't stir it all up in him all over again. There was a song in it that moved him about scheming sheriffs and laughing with Little John and escaping and finally making it. *Oo-de-lally, golly what a day.*

"Escape and finally making it," he said.

"Daddy?" Marionette said.

But he was off contemplating more than nothing.

That night he went downstairs to the basement and worked at his desk for a while like he would sometimes. He pulled out the old drawings he'd made of his castle. Another one he'd made of updates to that fort. He pulled out dream houses and subdivisions and even a monastery he'd thought about building, one with a brewery and a farm in the back.

Would such things ever be?

"Oh yes," The Good Lord said.

"How so, Good Lord?" Remmy asked.

"Go deeper," The Good Lord said.

"What's that supposed to mean?" The only thing deeper was the bomb shelter that he built in case of Russians. They used it for storage, nowadays.

Well? Why not?

Well he opened that hatch and went down in there and saw his trunk full of toys. Opened the trunk. First thing he saw in that sucker's Lincoln Logs. They took his mind to that wagon when he'd been five or six. He remembered waking before dawn to the cry of the cock and waking again a little later in Daddy John's horsedrawn milkcart. Daddy John would deliver milkjugs on one side of the street and Remmy'd swap them out on the other.

Down one of the streets there in Odin, Remmy started playing in the milk cart with some of the Lincoln Logs he'd brought along. He didn't have many, and they were pretty old, but he'd found them in the road on one of these trips, and it helped pass the time to imagine he was building a castle where all his friends and family could stay forever. Where they could prank a tyrant with an impune prank and then tell jokes with one another.

Those same Lincoln Logs he now held in his hands as an old father himself, older than Daddy John had been on that day, even, holding and contemplating.

In that cart, he had looked up suddenly and saw no one. Not a soul around him. Not Daddy John. Not the neighbors. Not any horses in the street. His dad had left him in the milk cart — had left him in the great wide world — alone. And had left him for work rather than for some other kind of play.

That's when the Wickaninnish blood first triggered — the predawn dark, the wandering thoughts about snakes in bags and bagless snakes and snakeless bags. He looked and saw no one in the canoe in front of him, so to speak. He looked and saw himself alone in the wilderness with nothing but an empty stomach and stones and heights and the devil to keep him company. He wished he had a bow in case of intruders. He wished he had a traveling companion or a guard.

He needed the shelter of his father. That or a castle wall.

As a five or six year old alone in that cart, Remmy had forgotten, for a moment, the consistency of his father's returning when he'd been gone in days past and only felt himself alone and working. He got out of the cart, panicked, and began wandering through Odin. He searched. He hunted. He chased. He was looking for people or the evidence that people had been there — the fog left behind after a car had passed, the hot shit of another horse that had just passed through, the cry of a crow scared out of someone's corn field, the smell of his father's milk cart. It was morning. There should have been people.

But people there were not: all sun and no stars, so to speak. Compassless.

He went door-to-door down the street as a man with short term memory loss might have done, looking for milkjugs that weren't there because he was no longer in their part of town now and looking too for people who still slept in many cases. Tufts of grass and gravel roads and muddy ruts and way more bright green and grey barked trees with old growth than in later years. Even still, the town was near deserted, and though he heard the milkcart and their horse in his mind's ear, when he came to the place where he thought they had been, there they were not there either.

So he walked home, which was quite a ways for six-year-old legs. Quite a long ways. He recalled thinking it was his journey to the end of the world, and there at the end he'd find his father dead. But he walked in and saw his dad standing there worried and said, "Well for a builder you sure give up easy on Lincoln logs." He'd laughed at his own five-year-old joke. But he'd stopped cause his father was red-faced now that he'd arrived.

"Where the hell have you been boy?" Daddy John David asked him.

"Looking for you, Daddy," Remmy said. "I got lost."

"Looking for me? Why didn't you stay in the milk cart? I was around the corner dropping off like we always do. Why didn't you stay, you stupid little boy? You left our horse alone for the horseswipes!"

"The what?"

"People't steal horses!"

Remmy was crying pretty hard by this point. "Wandered

all over trying to find you. You left me, dad! You left me for your work!"

"I was on t'other side of the street! How'd you miss me?"

"How'd you miss *me*, dad? How could you let me go?"

"I was working! You should have stayed put!"

"You shoulda been spending your time with me while I was playing castle."

"Thought I was," Daddy John said.

"I wanted a little rest with you."

"We can play while we're working. We can whistle. We can love our work, boy, and show our love for one another by working our asses off for the family."

"That's not how a feast works, Daddy John. That's not how the knights would feast. That's not a holiday."

Daddy John had sighed and started picking the mud off of his boots with a busted wheel spoke.

"Well least you could have done is let me work with you, Daddy John," Remmy said. "I was left in bad shape."

"Bad shape," Daddy John said. "Like Bloody Williamson. Good God what have I become?"

"What's Bloody Williamson?"

Daddy John turned to him and looked at him square. "Bloody Williamson is the worst thing a boss can do to his workers and Bloody Williamson is the worst thing workers can do to their boss."

"And we done did that to each other?"

"Yes. And I'm so sorry."

They were both crying by that point. "I'm sorry too, Daddy John. More sorry than you'll ever know."

Daddy John laughed, through his tears. It was messy. "You'd be surprised, boy." He picked at his boots.

There in the bomb shelter as a grown man, Remmy rolled the rough wood of those Lincoln Logs, spared as if from a wreck. Spared down here in the shelter. They might have been the pillars that Samson shoved. They might have come from them cedars of Lebanon that The Good Lord himself had split to humble King David. Any log like that spared from any wreck — even the wreck of a milk cart — took on a sort of mythic mold in his mind simply because it might have been lost forever.

Like him.

And he'd been trying to catch a break ever since. First with John David and then, when that'd failed, working up to catch a break with his family. You either built Camelot or you didn't.

But how was he supposed to build it with a thieving company man like Jim Johnstone right here in their midst? How's Robin Hood pull one over on that old sheriff?

He shuffled through the trunk again. He saw his old tin toys, some of them windup. He saw an old set of marbles: he'd once been the finest marble shooter this side of the county line and he'd taken more than one shooter from the boys he'd beaten.

There in the bottom were some old books. Most of them comic books about detectives and war and cowboys.

One of them was a comic about medieval times. He flipped through it and then saw his old longbow. He looked over in the corner and thought he could go hunting again.

But that wasn't it. That wasn't *deeper*. Flipping back through the book he saw jousting with horses. Another piece, but that wasn't quite *go deeper* either.

He turned the page and it showed another castle, this

one under siege. And the castle fended off intruders with a trebuchet.

Where had he heard that word?

He climbed out of the bomb shelter and went back to his desk and got out his old books from carpentry school. He started re-reading the books on carpentry and recalled scaffolding or house building jobs here and there with Daddy John back when his father'd started taking him along. He'd read one of those novels in the bomb shelter and then read a book on building the studs of a house and then he'd go to class and repeat. That's how he'd done it. He'd learned about what tools they used to build the pyramid and what angles they use in the crenellations of castles and how that works for the gable of a big house. Somewhere in there, he learned about how a trebuchet works — a word he pronounced "treb-u-chet" for the same reason his neighbors pronounced "Cairo" Illinois "Cay-ro" like that nasty syrup. That trebuchet had circles and squares and triangles all on it — its base was isosceles and its arm was ram-rod straight and it had a little sling in it just like the joke about the guy with the sock. He'd worked that idea over and over and drew so many different versions of it because it had so many trigonometry thoughts built into it. That was the only thing he hadn't yet applied all that well yet on account of it being something out of the old comic books he'd read as a boy. But most of the other stuff he had used on the different building projects Daddy John took him to. Southern Illinois wasn't a vacation with Daddy John. And she wasn't paradise or the promised land.

But he could defend her from brigands.

V. Rondo

...All that is Earth has once been sky;
Down from the sun of old she came,
Or from some star that travelled by
Too close to his entangling flame...

— Jack Lewis

For the Great Gaels of Ireland
are the men that God made mad
for all their wars are merry
and all their songs are sad

— G.K. Chesterton

WILSON REMUS

1974

SOON AFTER THE new year when the work went slower than normal, he rounded up everyone he'd moved into that neighborhood. And some of the ones that hadn't moved but who might have moved had they had the money to come and join the sharecropper side of the street. He called himself a conclave. He called Pete Taylor down. He called Ryan. He called John David his father. He called them all. All his Merry Men.

And they came.

His living room was packed to the brim, and Pete Taylor's propped up on the counter in the other room, peeking out with his big old head.

"Well?" Hellman asked. "What's this all about, Remmy?"

"This is safe space," he said, looking them all in the eye. "If anyone in here ain't okay with that — if any of you even thinks for a second they might be a gossip and can't be trusted with the truth of things, you can leave now without no shame in it."

Beth said, "Well Remmy, I'll be downstairs working on my projects, then."

The boys chuckled.

Norm stood. "I'll join you Beth."

"Good man," Remmy said and nodded. Two other gossips went. He waited until they was gone. Then he said, "We are here to talk about Texarco."

The room stirred a bit.

"I'll begin," Remmy said. "They did not keep to regulation with their derricks and when a storm came through, one of them derricks of theirs fell and busted open on our land and it ruined and contaminated and adulterated and befouled all our wells."

Several of the boys nodded and added details to that part of the story.

"Remmy had to dig seven wells, all dirty," Pete said.

"That's why I moved out here," Remmy said.

Hellman said, "They stole the oil right out from underneath me. Bought out the rights to all the land around me and sucked her dry and it ruined my well too *plus* I didn't get no money from it, neither."

Pete Taylor said, "They made my family relocate. I can't see you all as much. But they gave me a job, so that's kind of a wash."

There was a beat. Several seemed to be thinking about *gainful* employment.

And measuring gains against losses, they went to talking again:

"Ruined two years of my farm's corn growth when their

pipe burst," Bullhorn said. Bullhorn was called that because that was his Native American name, but also because he could be loud when he wanted to be. "On land supposed to be protected! Land of my people!"

Some of them shouted their agreement.

Remmy's face fell. And he felt high tide in his eyes.

Beth's Daddy, who never said a word, spoke up just then and said, softly, "They may be ruining my pension."

"What?" Remmy said.

"I don't want to talk about it right now," Beth's Daddy said, "but they might've screwed me and Beth's momma out of a pension."

"Me too," Daddy John said.

Remmy said, "Daddy John, you didn't—"

"I don't want to make this about me," John David said. "I just thought it would help to know that."

They went silent. Because what could a man say to that? You work loyal to a company for so, so many long and weary years and then they just go and cut the legs right out from under you and sell them too for the meat and stab you with your own femur once they've sharpened it up a bit.

Someone patted John David on the back somewhere in there.

"What do you have in mind?" Pete Taylor asked.

"I think of you boys like my Merry Men," Remmy said.

"Merry?" Sinclair asked. "Like Merry Christmas?"

"Sure," Remmy said. "You're the friends who make me happy."

"That make you Robin Hood?" Hellman asked.

"I don't know," Remmy said. "But we might as well rob from them rich thieves and give to the poor."

"How exactly are we going to do that?" Daddy John asked. He was getting old, you see.

Remmy told them of the first faint edges of his plan. The number of wacky things grew and grew just like in *Wacky Wednesday.* They couldn't escape them. And once every wacky thing had been counted, they all knew it would end soon. They were his Merry Men. This was his Camelot.

For better or worse, this would be *their* once-in-a-lifetime, all-out, world's biggest prank. It was time.

WILSON REMUS

1975

THEY'D ALL GONE into The Shed together and measured every piece of that makeshift shelving and logged it like stock boys at a grocer's. And once they'd done that, turns out they had way more than they thought.

On his way out of The Shed, he saw a large object at the back under a big old tarp and he wondered at it.

Then Remmy went to work designing the thing just like he'd done as a boy, keeping track of the different sized logs and how many he'd need to build his work. He went to drafting down in that basement.

Somewhere in there, he went out to the land in the woods near his house and checked to see what was left of the fort. He needed some time to think. Some space. He also needed to know if Jim Johnstone'd taken it apart board by board and stone by stone. Or if some invading army of tycoons had besieged it.

Or if the earth had swallowed it whole...

There in the woods it stood, still and sure. That Johnstone asshole had lied out his teeth. He'd probably only bought it for the mineral rights and left it fallow or was waiting until the land built up and needed expanding like the rest of the neighborhoods.

Remmy went out there every day for four hours and did nothing but pray. He became half-monk in that time. He didn't shave. He didn't eat. He barely drank anything. You coulda told the little kids in town that he was out there wearing an old horsehair shirt and eating nothing but grasshoppers and honey, and they would have believed you and retold the story and added nasty teeth and claws and whatnot. But that wasn't really the way of things. He was just out there praying and hearing nothing whatsoever from The Good Lord. Hell he wore out the knees on his pants praying like that. You see, sometimes a mind's gotta figure stuff out on its lonesome. Sometimes you put too many incoming letters in a mailbox, it can't send no more out and ain't good for nothing but getting hit by a baseball bat out a car window going sixty. Remmy's head felt like that: like he'd been given too much, like he'd heard too much, so he did most of the talking that week and The Good Lord just sat and listened.

At the end of that week, Remmy needed to go out to the oil fields, and he decided to walk. Out there in his pitched tent near the fort, he measured his average stride. He did this by measuring his stride twenty times and taking the average. He wrote that number down. Then he went out to the fort and stepped into the glade near where it would go and he walked and counted and walked straight south by southwest into the heart of Texarco land.

After walking a mile straight through the woods

— docking a step or two if he had to detour around a tree — he came out into the fields with some of the largest derricks there were. He saw another half-mile ahead the small mobile home that held Jim Johnstone's office, the one that looked like a battle tent.

He walked into the office where he'd once nearly killed the man and asked him for his check. You see, Remmy'd been hired to build a couple of houses for Texarco straight, a small neighborhood they planned to house folks they paid on script, which is pretty much like getting meal rations in a military you was drafted into in the first place. Remmy doubted he would have done it had he known at the start what they planned on using them houses for. But he was in it now, and there was Jim Johnstone with that look on his face like he had bad news nobody else knew about and he'd only tell you once you begged him good and long.

Jim's signing that check and says, "You should really think about getting into oil, Remmy. As much property as you've dealt, I could go in with you on some mineral rights. I know just the place. Really cheap, good well, puts out on the regular."

"I don't want oil money," Remmy said, "but thank you, Jim. I'll just take my check and be on my way."

Jim smiled. "You can do a lot of things with money you can't do with your jokes and your charm and your stories, Mr. Broganer. You can own the world as a company man. You'd make a great company man."

"I don't know what you're selling, Jim Johnstone, but ain't nobody in this town calls me Mr. Broganer except Miss Witt, and she's been a widow in a lonely house out at the edge of town for a couple decades now."

"Oh quit it with your theatrics, Remmy. Let's talk as men talk. Take Terry over here."

His secretary perked up. She was a mouse of a thing. A gorgeous young woman once whose life and light had leaked out of her working for such a man as this for so many years. There was no fight left in her for anything. You could have bombed a poor Iranian village and she barely would have flinched. That's what comes of working for a man like that for as long as she had and no one never standing up to him.

"Yes sir?" she asked.

"Terry could you go grab my folders out of the car?" Jim asked. "That new study out of the north?"

She nodded, got up and walked through the door and down the steps of the mobile control center.

He pointed to her backside. "Take her for example. Sweet little thing. Good breasts. Nice ass. When you have so much money like this, money gives you power and honor, you see? Cause you can buy up land and things and trade them for favors, and favors give you all sorts of things." He wagged his finger at where she'd been.

"I don't think I need that kind of money." Three minutes in, and Remmy was already tired of listening to this old fool, but he figured he'd get more out of him as a friend than a foe. "But what do you have in mind?"

"See there, that's all I ever wanted out of you," Jim said. "I was thinking we could buy up those old coal mines in the south and expand. You could sell your business, buy up some of the coal say… around Williamson County area. We'd split the oil and the coal earnings and then buy up some of the timber. The coal and oil will run their course anyways, and timber and farmland are gonna be the future. We'll want in on it before everyone else."

"Okay."

"Okay you'll do it?"

"Okay, I'm listening."

"Well I figure you can sell your real estate and business and make it happen pretty quick. You'd have to take out a loan, but hell, as quick a profit as we'd turn and as cheap as labor's getting, we'd be able to turn it around. And then you could have as much money as me and we could be better than friends."

"I'm not gonna be your lover," Remmy said. "That's taking favors a little too far."

"Nah, there's something better than friends and better than lovers."

"What's that?"

"Business partners."

Remmy tried very hard not to laugh. It took some doing.

Jim walked up to the window and stared out at Terry bent over and rummaging through the car. "Then we can group our favors together and we'll own half the state. We could run for office. We could have any girl we wanted. I'll even share Terry."

"Share Terry…"

"Now I know I'm a downright mean old cruel son-of-a-bitching scoundrel sometimes," Jim said, "but don't pretend like you're all innocent and don't know what I mean, Remmy. She was down at Indiana Beach. I tried to beguile her with my charms, but I couldn't do it."

"She's married, Jim."

"So? You just gotta snap to them, hands to hips, and

either it clicks or they repel you away. When you're an oil man or a coal, it usually clicks."

Remmy couldn't fake it no more. He couldn't even kettle pitcher his way out of this kind of babble, not even for the sake of Jim's soul. And it was showing on his face.

"Don't act self-righteous," Jim said. "All of us men are the same."

"Jim," Remmy said, "You made up my mind."

"You'll join in with me?" Jim smiled. "I knew there was a coal man in you. It's about time we remade Williamson County in our image."

"No, you made up my mind about something else. I been looking for a sign, and I think you just gave it to me."

"What's that?" Jim asked.

Remmy just grimaced. He looked at Jim's corncob pipes.

"You want to try one?" Jim said.

Remmy recalled his grandad's cigarettes, the ones he put in a humidor every night so that *they'll keep, the'll keep*, the ones he carried around in his breast pocket just in case. It was another sign, had to be. "Sure. But only if you try one of mine."

Jim made him a pipe and passed the peace.

In return, Remmy handed him a shit-filled cigarette, which passed the war.

They smoked.

"My this is great," Jim said. "What's in this tobacco?"

"Alfalfa," Remmy said.

They smoked. Remmy never smoked, but he figured where there's smoke there's fire, so why not smoke here at the end of everything that ever was?

"Oh it doesn't matter," Jim said. Bastard loved listening to himself. "Point is, money gets you favors and favors gets you honor and honor gets you anything you want. You can hire Mexicans for cheaper than the white folk like I've done, and then run them right back out of the state once they get a little power and give the jobs back to the chastened white folk for half the cost. That's all it takes. All you gotta do to get white trash to let you walk all over them is teach them, little by little, that the poorest and worst white man's still better than the best colored man on earth. Hell that's why Bellhammer and Salem and Carlyle can take so much from Centralia. Did you think it was industry? Nah. We should have been a slave state but that damned Archbishop Ritter done ruined everything with that excommunication bullshit over integration back in the forties. Money gets you favors, Remmy, and favors gets you honor and honor gets you anything. You're really missing out on this oil deal. I need a good friend like you. Someone I can trust."

Jim looked up. The bags under his eyes looked lonely and alone. Looking for trust to the man who just handed him a shit cigarette.

"Jim Johstone?"

"Yes, Remmy," he said, tearing off the check.

"You are many things, but honorable will never be one of them."

He took the check and looked down at the desk and saw a ledger book with all kinds of names in it — names of friends and relatives and folk from Bloody Williamson. People't owed old Jim Johnstone half their lives. And he smiled as if he'd just passed a great test. He walked out the door and closed it behind him, Jim still inside smoking shit.

Remmy looked to Terry who had stood up from the car. He said, "I'm sorry he treats you like this. Anytime you want to quit and come work for me, there's a place for you."

Terry said, "I think that would be the best thing to ever happen, Mr. Broganer."

"It's Remmy. My wife's Beth. And we start at five in the morning and close early."

"Yes sir. And you'll be wanting to read this over."

She handed him Jim's folders. Inside was a report entitled *Arctic Projections of Worldwide Warming and their Effects on Arctic Drilling*. He grabbed that and put it in his satchel.

The door to the mobile home swung open. Jim shook his head, marched down the steps, spat in their general direction. "Terry. Get back inside now."

"No," she said.

"I said get your ass in the trailer."

"And I said no."

Remmy watched, his hammering muscles tensed.

"Now," Jim said.

"I will not. I quit."

"What?" Jim asked. "Why?"

"Cause you're a downright mean old cruel son-of-a-bitching scoundrel," she said.

Jim looked hurt. He turned and started marching back into the mobile home.

"I'm going to work for Remmy," Terry said.

"Stealing my cattle now, Remmy?" Jim called over his shoulder. "Nothing but a damn horseswipe after all."

That hurt not just on account of calling a woman livestock but on account of what Daddy John had said about leaving the milk wagon, but Remmy didn't let it show. Tried anyways.

"You have any supper?" Terry said to him. "I don't have any money."

"Sure. Beth always makes plenty."

So she came to dinner and Remmy helped her get a job doing the books with Beth and, my, did those two girls ever talk. Made fun of Remmy, mostly, and all the trouble he got into.

Remmy started to read up on *Arctic Projections of Worldwide Warming and their Effects on Arctic Drilling.* Turns out Texarco had hired a bunch of them old scientists to make themselves a voyage up north to see what they could see about drilling in the frozen wasteland of Alaska. Or at least they called it a wasteland. Turns out a great many nice and good and creative people'd been living long and fulfilling lives up there for years and years. But Remmy guessed that didn't count as human in the mind of Texarco on account of them all being in the way of their pipes and derricks just like the folk in Southern Illinois.

Anyways, Remmy was a good reader, even though he didn't talk none about the things he learned in the fictions and nonfictions the world printed on its presses. And he read that several hundred page report and skipped the math and then read it again and tried to look at the math and then read it again and he began to get some of the math, and the best he could explain it all to Beth was something like this:

"You know when you get in the shower in the morning and close the door?" he asked.

"Yup," she said.

"You know how it gets steamy?" he asked.

"Yup," she said. "I like it steamy."

"I know you do," he said. "But it gets so hot."

"That's what I like."

"But you like opening the door," he said.

"Exactly. Hot and then cold. The rhythm of it. Like summer and winter."

"Well what if you couldn't open the door?" he asked. "Ever?"

"Then I suppose that'd be awful," she said.

"Well remember that time I brought you in that glass of ice water while you was steaming and all the ice melted and half the glass dried up?"

"Yup," she said. "You had to pour another."

"Well what if you could only get all your water and air for your shower out of the bathroom? What if your water was just a little pond of ice water on the floor like that glass? And what if the more you pulled water off the floor and put it in the air like steam, the more it heated up all that ice water on the floor until it melted and turned to steam? And what if you did that so long suddenly you still had that water heater on, heating up the air, but no more water to fill it?"

"I'd probably burn down the whole damn house," she said. "But that wouldn't happen. For one, that's water vapor, and if it's that saturated it's gonna be pretty hard to evaporate any water in it. And besides, you're omitting the science of displacement. I remember displacement from science. You have an ice cube in a glass of water and it displaces so much and then it melts and it displaces the same amount.

Don't matter how much of it melts, it'll still have the same amount of water."

Because, you see, Beth knew a few things, too.

"Oh Beth, it don't work that way. Most of this ice is up on Greenland and Canada and Russia and whatnot. The ice ain't in the *glass* yet. As it melts, it's adding more and more water to the seas."

"That won't happen, Remmy."

"It is happening." He waved the study. "That's what they're doing to Sister Earth, you see."

"Sister Earth?"

"That's what St. Francis called her, ain't it? Them oil boys are *counting* on it even. Counting on all them ice shelves and glaciers and whatnot melting just like my ice block and your glass of ice water did. And as it melts up in Canada and Alaska they're gonna drill some more oil for us to burn. And as it burns, more ice'll melt and reveal more land and, hell, they'll just go on a drilling all over again. They've known about this for years and they're starting to use the knowledge to their advantage to get us to use more gas so that the ice melts and they can drill for more oil up in Alaska. That's why they're spending so much on lobbying the Army Corp to build them roads: trying to get us to use more cars and less trains. Bet that's why they even invented them Michelin Stars: to get us to travel more in the cars."

"But what about the rest of the house?" she said.

"They'll just watch it burn." He slumped in his chair and thought of poor old Terry who'd had to work for such a man for so long. "You'd think they'd realize that burning the world to make a buck today ain't gonna make it very easy to make a buck tomorrow. Among other things." He thought of Toby and that broken bottle of ketchup. "Like babies."

"Oh Remmy," Beth said, "don't you see? Those nasty old men count on being long dead before that."

"That's about the most selfish thing I ever heard, selling ten-year poison to your grandkids cause you know you're going to die this month and you want money for gambling and drinking and smoking. Who sells out humanity for the price of a little bit of land?"

"Well Remmy, it's not like there's only one Judas. Didn't Esau do it too?"

It was not having consequences was their problem. Like little boys who didn't have a daddy to spank them none.

And now there would be consequences. *Arctic Projections of Worldwide Warming and their Effects on Arctic Drilling.* He could go to the papers and to WJBD, once he'd got their atten- tion. It was at least his duty to defend Sister Earth — to defend Little Egypt, his paradise — from brigands. Castles don't have walls for nothing, after all. He could save her. He could save Sister Earth if he jested big enough, loud enough, strong enough. This wasn't about some joke as the be-all, end-all of his life no more. It wasn't just some joke or prank. Grandad Patrick had been right: all good Dempseys do jail time at one point or another, but he could always introduce the report as leverage. And if he timed it right… Ten years, probably. Ten years's a nice round number: a man can change a lot in ten years. A company that hurt its workers'd grown big enough to hurt the whole of earth.

He sang softly to himself: *This is my Father's world and to my listening ears all nature sings and round me rings the music of the spheres. This is my This is my Father's world, He shines in all that's fair; In the rustling grass I hear Him pass; He speaks to me everywhere. This is my Father's world. O let me ne'er forget that though the wrong seems oft so strong, God is the ruler yet. This is my Father's world: why should my heart be sad? The Good Lord's King; let the heavens ring. God reigns; let the earth be glad.*

It grew quiet as only it can when one sings a hymn on one's lonesome. Pascal says all men's miseries derive from not being able to sit in a quiet room alone. And Augustine said that he who sings prays twice. Well Remmy'd prayed twice and now he sat in a quiet room alone:

. . .

. . .

. . .

He chose. Then acted without any doubt or delay. First he went and got the second money sack that he'd saved, not the Merry Men fund, but the one Grandad Patrick had passed on to him IN CASE OF BLOODY WILLIAMSON full of FDR bonds, heirloom seeds, cash, and gold. This he gave to Beth in case Jim got desperate and gave her the name of an attorney who understood what's at stake. He told her she might need to run the books for a time in case Jim got desperate but he didn't explain much more or mention bail money.

Then he went back to the fort to pray and draw some more of them drawings and measured and measured. That's what they always say in carpentry school: measure twice, cut once. He measured the iron. He measured the distance. He measured the fort. He measured and measured and measured and found Texarco lacking. And then he measured and measured his plan and found it full. Just over a mile and a half from that fort, he uncovered the full length of the throat of Texarco that'd come to swallow them all alive. In Southern Illinois, it turns out carpenters and barbers did have something in common after all: the way they came at them stretched out, oily throats.

Measure twice. Cut once.

WILSON REMUS

1976

ABOUT NINETEEN-SEVENTY-SIX REMMY said in later years that they was just a normal two-kid family coasting through life.

Bull.

Shit.

That night, he fought a groundhog underneath the house. Technically they had a full basement, but it didn't cover the entire floor plan. Some of the house — the newer part — had a crawlspace. And it was somewhere down in that crawlspace Remmy heard a rumbling at around two in the morning.

Beth said, "Oh God Almighty, the earth is swallowing the house!"

"Like hell." Remmy stirred awake. "I was dreaming about tournaments."

"No time to talk about basketball, there's a monster under there," she said.

"Not basketball, knights. Like Sir Lancelot used to fight in. Horseback with lances and—"

"Remmy!"

"I'm going."

He got to going.

Underneath that fiddleback infested crawlspace, he used an old deer spotlight, you know like hunters and the police both'll use'n a dark woods, one for the living and the other for the dead. At first, all Remmy saw was fiddlebacks and Remmy was scared. A kid in his youth'd gotten bitten by one of those one time, and it was like death itself started to rot that boy's arm off, spreading like the plague, spreading like how rot'll spread in a tree trunk struck by lightning, until it killed that boy. He moved past slow, scared for his life and his arms. Shining that light further, he saw one of them groundhogs had started digging on the northside.

He got over there and didn't know if the thing had dug under it, under the whole house, under the wall. He looked up there and saw gnaw marks on the floor and a bit of light where they'd started to get through to his brand new family room, and oh was he ever mad. But he wouldn't stick his hand under that hole to see. He wasn't about to have his hand eaten off by a groundhog.

No sir.

He crawled back past all them fiddleback nests with struggle and sweat and fear and went upstairs and looked for a hand mirror but he couldn't find one for the life of him. He looked everywhere he thought Beth might have in there. He couldn't get a mirror.

"Mom," he said to his wife Beth.

She'd gone back to sleep.

"Hippo Shit," he called her.

"What?" she asked and shot up in bed. "Did you get it?"

"No."

"Well try—"

"Do you have a hand mirror?"

"What do I need one of those for? Leena does my hair."

"Dammit."

"What?" she asked. "Is it a curse?"

"Not quite," he said.

"Then why you cursing?"

"Cause dammit, Beth, sometimes I just like to at two in the morning when beasts of the netherworld have risen to tunnel my house!"

She snorted out a sigh.

He said, "Get under there with the flashlight for me and stick your hand down in there."

"You must be crazy," she said.

"I'm saner than stainless steel."

"No, you're crazy, cause only a crazy man wouldn't have the spine to shoulder his own curse like Adam did for Eve, asking his wife to grab hold of some monster in some pit cause you got no spine yourself."

Bren was standing in the doorway eating one of Beth's peanut butter cookies, munching loudly and grinning.

"What the hell you want, son?" Remmy said.

Bren munched loudly. "Oh, go on. Don't let me stop you."

"You don't understand, son, stay out of it."

Bren said, "Groundhog's under the house. What's to understand?"

"Yeah well then you go under there with that big

flashlight and stick your hand down in there and see if it's there and you can pull it out."

"I ain't going noodling for no groundhog."

"Then shut up and go to bed," Remmy said, "cause it sounds like this whole house is as cowed as me."

"What are you gonna do?" Beth asked.

Remmy thought for awhile.

Bren munched his cookie.

"They any good?" Beth asked.

"Ain't none better, you know that," Bren said.

Beth tucked her chin and grinned like she did when she'd fished for a compliment and got what she wanted but still hadn't expected on account of that insecurity her own daddy'd put in her.

Remmy said, "Holsapple's got a pistol. .32 caliber."

Bren grinned wider. "There you go."

"I'm leaving," she said, "You'll be shooting through the floor and I ain't about to die while lying down in bed." She got out of bed and went to the closet and started pulling out clothes to get ready.

"I'll set up the lawn chairs." Bren went out and started both that and a fire and got stuff for s'mores. The lawn chairs weren't so white and weren't so new in those days.

Remmy went over to Holsapple's. It's no soul's hour by that point. He starts banging on the door. Jerry comes downstairs and opens the door and shout-whispers in that hoarse way that mad men trying to stay quiet for sleeping wives who wear the pants will do. "*Remmy, what in God's name?*"

"Don't use his name in vain," Remmy said at a normal volume.

"*Whatever this is, it has nothing to do with vanity.*"

"You need a new one?" Remmy said. "I build nice vanities. I'm a carpenter, you know, J—"

"WHAT DO YOU WANT?" Jerry didn't do too well low on sleep, you see.

"Jerry?" Mrs. Holsapple called down.

"Oh God," he said. "See what you went and did?"

"I didn't shout none," Remmy said. "I wasn't the one shouting."

Mrs. Holsapple came down in her nightie. She reminded Remmy of Joe's wife standing half naked on the porch after the oil derrick fell down from heaven in the middle of that tornado all them years before. "Evening, Remmy."

"Evening," he said. "Sorry, Momma Holsapple, but I need to borrow your pistol."

She gasped.

"I have an intruder," Remmy said.

She gasped again.

Jerry suddenly sobered. "Come over here. Come over here quick."

Now. Every man in Southern Illinois has a closet or a trunk where he keeps the guns in case of armed rebellion like Bloody Williamson. Jerry went to his. One of them was that sawed off ten-gage with the double hair trigger Remmy used to own back when the boys tried to T.P. his house and he'd had them teach him how to throw them rolls up in the boughs of them trees like bad Halloween tinsel and the cops had come and tried to fine him for hitting his own yard.

Remmy'd got rid of that shotgun cause it was double barrel for one and cause it was a ten gage for another and cause it was sawed off for a third and cause, for a fourth, when you combined all that with a loose and faulty hair trigger on the back and you squeezed the front trigger, the back trigger would pull off a shot as well and you'd bruise your shoulder accidentally shooting both barrels at the same time. And also cause for a fifth if you bumped it just right, that hair trigger would go off all on its own, and when it went off, it'd make the front one go like the time with the tree and T.P.ing. So it was both barrels either way, and he didn't want that, no sir, so he sold it cheap hoping Jerry'd use it for parts cause Jerry Holsapple liked his guns.

THAT sawed off faulty ten-gauge was leaning in the closet full of buckshot while Jerry's in there rooting around for his .32 caliber pistol. Well he bumped that faulty shotgun and it fell down on the floor and the faulty trigger got squeezed and both barrels went off. When they went off both boys jumped up like alley cats when they hear a wolf set to howling. That shotgun blew a three-foot-round hole clean through Jerry's outer wall and luckily there's an old elm tree not too far away and it took most of the shot but a couple of them little bb's went and shattered the big bay window next door at Jim Johnstone's house. Then his lights are on and he's out in his yard cussing and swearing about hooligans breaking into his house.

Some people get really tight and tense when they're scared. Some people fight and some people flee. But some people are more tightly wound than others. Some people spend so much time running around and fighting and fleeing that when something *really* bad happens, they just go all slack like they took a morphine shot. "How did that just happen?" Mrs. Holsapple asked, serene as hot chamomile tea.

"If it can happen," Jerry said, "it'll happen to Remmy."

"That sounds dark," she said.

"Guy named Moore wrote a law about it," he said. "What can happen'll happen."

"But that's both the bad and the good," Remmy said. "Not just the bad that can happen will happen. The good that can happen will too. And if you happen upon bad things like tonight and in that moment you choose to happen your good on them bad things, more good than bad'll happen in the end."

"I don't think that makes much sense," Mrs. Holsapple murmured.

"Sure it does," Jerry started, "It—"

"Shhh, shhh," Mrs. Holsapple said, "Let's go cheer on Remmy."

Remmy loaded up the pistol and marched out, the last knight in the world striding face-first into the dark of the last night of the world.

By then the whole neighborhood is standing around Bren's fire in the front yard, cooking smores and talking about baseball and sipping hot toddies and watching and cheering — well everyone except some of the Texarco folk, who were frightened back by Jim Johnstone — as Remmy crawls under his house with a pistol to chase the monster out.

He crawled alongside the fiddlebacks and he stopped and looked at them, one of them dangling like it was about to drop on his nose. And he felt his heart set to pattering.

The Good Lord said, "Remmy, have you seen the fiddleback giving birth to its young?"

"Can't say as I have, Lord."

He felt the Good Lord smiling. Above him that fiddleback started attaching an egg sac to its little cobweb. And then Remmy saw all them egg sacks up above that it had already laid. You know that each egg sack of a fiddleback has somewhere between thirty-one and three hundred baby spiders inside? Well they started hatching just then, and the floor of his house which was now the roof over his head started boiling with baby spiders. Just boiling and roiling.

Remmy was sweating. He said, "But people used to worship spiders, Lord."

"They're not gods. I am."

"I want to squash them all."

"They're not demons either. They have my spirit of life in them, Remmy."

Remmy looked at that spider again threatening to fall upon him. "Sister Fiddleback," he said. "I pray you be patient with me as I pass."

He moved through unmolested.

There at the northside, he shined that light down into that hellhole again and saw them yellow eyes. Plenty of them. He pointed his gun down in there.

The Good Lord said, "Remmy?"

Remmy's hand shook on the pistol.

"Have you ever seen how pretty a groundhog can be when it digs?"

"No, Lord. Now's not the time."

"Seeing as how you're aiming that death cannon at it, I think it high time. They look like they're swimming, Remmy. I made'm thataways."

Remmy aimed his gun.

"They're just doing how I made them to do."

"Digging through my house?"

"Your house I gave you."

"I built it," Remmy said.

"With hands I gave you, with a mind that received my gift of consciousness, with lungs that's got my breath in'm, and with wood that came from the stock of trees my spirit told *grow*. Tell me: which of these things did not come from the very dirt these groundhogs live and move in?"

Remmy didn't have no answer for that. He wanted to ignore The Good Lord and put a bullet in these things and go to bed, specially with half the town cheering him on just outside the cinderblocks.

"Remmy?"

"Aww shit," he said. "Can I at least wound them a little?"

"What you do is your choosing, but they're as much mine as you are. You'd do well to remember that."

...

There was a pause long enough to chew and swallow a bit of burger.

...

"I'm gonna wound them," Remmy said.

He pulled off a shot and missed all four of them, and his ears went ringing and them crazy ass groundhogs came out and clawed at his face and arms and — scared out of their minds — they went to running all over that crawl space, and he started shooting at them left and right and missed — my God he was a horrible shot with a .32 caliber pistol — and he shot holes all through his family room floor, which hit pictures and books and things up above, not least was the grandfather clock John David had made them as an anniversary present from the benches of pews from Young's Chapel where they'd been married. Somehow, some way, Remmy reloaded and wounded two, and they ran off and never came back.

He came out bloodied and beaten and as wounded as the groundhogs.

"Oh my God," Beth said. "Let's fix you up."

The men cheered.

The women covered their mouths.

The children looked at his bloody face in horror.

Remmy was down for days and then had to get to work repairing both Jerry's wall and Jim Johnstone's window.

He kept meeting weekly with the Merry Men between jobs and the crowd had grown a bit and the prank got more and more elaborate. Jim Johnson of Texarco tried to invite himself inside one time. He thought it was a party.

That fall was the Marion County Fair. Big fair. Nice fair. They did tractor pulls and horse races and beauty pageants and what not. You know, the kind where you enter both your fattest pig and your oldest daughter to win blue ribbons for prettiest and then bring you home money from whatever suitor the grand prize attracted.

He sat in the grandstands watching the demolition derby. A couple of his nephews competed that year along with his buddy Virgil, who won most years, and he watched them smash steel-reinforced Buicks and Chevys into one another over and again. So it was seven of our relatives, give or take. He watched one Dodge come out and power through and hammer into different cars — boom, boom, bam. Then it died.

"Owp," he said. "That's it for him. Those things won't start back up."

Sure enough, it never did get back up.

They smashed and they smashed, coming at each other like...

...like knights used to do in his comic books.

Except it wasn't no horses here. Other than the horses, that is.

It was a strippeddown Buick with a bunch of spray paint where the horse blanket should go and a grill and hood ornament right where the lance should have lead out front. Knights and their jousting.

Texarco had a gas line that the guys worked down there at the oil fields ran from the highway. They always had gas stoves — didn't have to pay for the gas or anything. They ran a pipe down this road that Remmy lived on, about a quarter of a mile, piped into all of their houses.

It didn't pipe into Remmy's because he didn't work for Texarco.

Well years later, they abandoned that gas line thing and people had to make other plans for heat. But that two-inch pipe's still laying in the ground and in 1976 said, "We want some of that," so Remmy asked to get some pipe.

Jim Johnstone said, "Take what you want, it's an eyesore, just barely covered with dirt like that."

So Remmy brought some up to Bellhammer and separated his garden from the gas and his compost heap too which smelled like shit. He dug up all that pipe. He built a swing set out of it, welded it together and painted it as black as the oil derrick shelving the boys were busy tearing down. The kids swung on that set, even older as they were, swinging with their dates, and the neighbor kids too.

Then birds got up inside there, nested up in that top pipe. Swallows. They'd chase him when he was mowing, dive-bombing after him like the Japanese did after the boats in the harbor back when he'd been six.

"They're so loud," Bethy'd say out there on the porch eating her chicken salad sandwich. "So awfully loud."

"What do you want me to do about it?" Remmy said, mouthful of his own.

"I don't know, fix it Mr. Fixer."

"I'm going to fix that." He swallowed.

Went into the garage.

Dug around where his M80'd been, the one that he'd used against old Pete Taylor's ankle, and found there on the bottom an old half stick of dynamite. He shoved that dynamite into the end of the pole, lit it with a long fuse, covered the hole with a big old piece of plywood and the birds shot out the other end like a cannon and hit the fence, dead.

"I think I'll take one piece," Bullhorn said. "But how am I gonna get it there?"

"You gonna drag it?" Remmy asked. "It'll wear out like a straw on a metal grinder and you'll have nothing left."

Seventeen-foot car and thirty-three-foot pipe, dragging it underneath the chassis, he did pretty good until he got to his house. He had to unload it and drag it over his yard with his lawnmower or something because the pipe was sticking in the drive like a car- sized pole vault.

Watching it go out like that and recalled that demo derby and the idea with the bird cannon all merged in Remmy's mind.

He went down and started drawing again.

They got the rest of that pipe pulled up and unloaded out in Virgil's yard. And they had a good old party because I guess Virgil had won again and wanted to celebrate, who knows? They rigged up them pipes on those cars like Bullhorn had and started running at each other like a bunch of fools and swirling in the mud and some nonsense. They got all tangled in Jim Hunter's fence and fifty head of cattle got loose like some African stampede you see on the T.V.

So the Hunters went after them on horseback, the ranchers woken up in the middle of the night to go at it, and Virgil and the Broganers and half the carpenters from Remmy's neighborhood chased after them cows in the demo derby cars with those lances on the front like some knights of the apocalypse and they got all but four cows caught plus those ones they accidentally ran straight through like god-sized kabobs.

So they paid Jim Hunter double for the four heads of cattle.

Slaughtered them.

And had a huge three-day feast where they invited everyone they could. Almost like a sacrifice before one of old King David's wars. Meanwhile Remmy modified all them pipes so you could get at them through the floor of them demo derby cars like an escape hatch, like the loader in a howitzer, and he put some of Virgil's tools in there and shot them out like some makeshift canon, using the dynamite for powder.

Put a hole in Virgil's steel barn. Like buckshot big enough to take down an oil pump, even the big ones owned by Texarco.

WILSON REMUS

1977

THAT WINTER, MARIONETTE got married to that pharmacist named Dean. Got married in some couple's backyard by the pool with a brown fence and floating candles.

Remmy didn't get much attention from folks on *that* day.

He knew it was his daughter's day.

Knew he wasn't supposed to say nothing.

Tell no stories.

Put up no spotlight.

Shoot no shotgun and wrastle no groomsmen.

Shoot dice with no preacher — that other one'd long died anyways. So he stayed quiet the whole time. The whole wedding, didn't say a word.

But that didn't keep him from printing out business cards to hand to people when they were asking what got him so tongue tied. They read:

I am the father of the bride.
Nobody's paying attention to me.
(*except the banker and the courthouse*)

After the wedding, the preacher came over to collect his due from the father of the bride and Remmy'd forgotten about that part, but Dean was already prepared. He walked in the house and brought out a five-gallon bucket full of pennies. "I had to save every last penny to pay the preacher," he said and slammed that bucket down like an old Irish pot of gold.

The preacher said, "You're joking. I need to roll each of those pennies for my payment?"

And Dean said, "We look like the joking type?"

And Remmy's tongue came loose. "Praise the Good Lord, I knew I liked this boy."

Dean and Mary snuck out of the reception and got to Lake of the Ozarks. Got to the gas station in the middle of town and had left the wallet in the front seat and locked the car. Remmy came and met them and picked the lock for them and got the wallet. Got to Carlyle and didn't have a suitcase and Dean said, "You don't need clothes for a honeymoon."

Remmy delivered them.

Didn't have her hot rollers.

"I will pay for your hair to get fixed," Dean said. "We're either going on our honeymoon or we're not."

He went to pharmacy school in St. Louis when they got married. Marionette worked at a bank. Dean worked when he could on weekends as a pharmacist. He worked as an assistant for the experience.

Good thing too.

Ain't nobody related to Remmy needed to be in Southern Illinois that year with what he had brewing. More than those that needed to be to execute what he had brewing, that is.

Cause he'd been brewing it since he'd been six years old.

Soon as the first thaw hit springtime, Remmy — with his wife's half of the torn-in-twain fifty-dollar-bill pinned to his lapel — took himself and his Merry Men out to the fort there on Jim Johnstone's land. They started rebuilding all those parts from that crashed oil derrick and fitting it to them big old tractor pulling chains them farmers used to get 'em out of the mud. Every time someone placed something wrong, Remmy knew and he'd point again to the blueprint and harp on them for not having good trigonometry. And then he'd fix it and show them how to fix it and then he'd be off to another angle. "A mile and a half," he said. "It's a mile and a half from here."

Nearby, a dump trucked idled with a big old load covered up with a tarp.

They rigged that up with a big old hammock made outta the tanned hides of those heifers of Jim Hunter's that they'd barbecued after accidentally killing them with the giant oil pipe demo derby car kabob skewers. Paul Tanner'd tried it out for his name's sake, tanning hides. They fitted it with combine wheels and a pullswitch like what you use on a grain car on a freight train. Rigged it up good and when they finished, they climbed up in that old ancient fort and looked out upon the massive steel oil derrick there in the grove, repurposed as if some god of war had found a better use for some god of greed's play toy, a redeemed piece of alien architecture, as if the angels had kicked their scaffolding for building the pearly gates over the edge of heaven in hopes it'd land on hell.

"It's awful," John David said. "I will have no more part in this."

"It's an awesome thing to look at," Remmy said.

They looked at one another.

Then they all stared at the massive steel oil derrick trebuchet, as tall as a water tower. John David left. So did a couple of other boys, even though they'd been done wrong by old Texarco. Most stayed. The ones that stayed lived up to their namesakes too. Bryan Fletcher fletched up some arrows for Remmy and the other bowmen using long goose feathers from the Chicago prank. Mike Smith had hammered up some crude swords. Even Dick Cooper had built some wine barrels and The Brewers had filled them with gallons of fresh dandelion wine. Old Jacob — the student who'd dressed up as Rooney's mail order bride — came out dressed in drag. Everyone of those old boys had come out and pitched in using skills their families had long forgotten.

They had a spotter exactly a mile and a half from there. Pete Taylor was dressed up in a tuxedo his wife had hand-stitched for a party, sitting in the driver's seat of Remmy's Magic Chevy, brought out of retirement. He was sitting, idling on the eastern side of the oil fields with a CB radio hooked up so he could run it two-way. Behind him idled a dozen demolition derby cars idling with those black pipe cannon spears sticking out from under the chassis, loaded with old scraps from the junkyard and a bunch of half sticks of dynamite. A couple of them were loaded with what they called "applesauce makers" — rotten apples from

Pickneyville. One wasn't powered by dynamite, but rather by that old double barreled, faulty-triggered shotgun.

On the other side they had a bunch of longbowmen in the hill sitting there with a phone hooked up on a closed circuit down the hill to Remmy's command tent, wired up with that old barbed wire spool he'd saved from the Barbed Wire Telephone Company.

Meanwhile, a little bit west and a mile and a half to the south, they loaded up the rusted down scrap engine block that weighed exactly three-hundred-thirty-three pounds into the sling.

Remmy started to sing a song he'd learned in Sunday School, "*Only a boy named David. Only a little sling.*"

They cleared the path of the oil derrick trebuchet and pulled the lever. It made a sickening dragging noise like some troll dragging his feet along the platform, but then the sling was up and it being a well-oiled machine, it went up and released that precisely weighed engine block and that thing flew through clean air.

"*Only a boy named David, but he could pray and sing.*"

It soared a mile and a half and landed twenty yards in front of the battle station mobile home of Jim Johnstone.

Jim Johnstone wasn't at work. Well he didn't have a secretary anymore and most of the other bosses were in their own fields or offices and the workers of this field? Well he'd sent everyone home for the day after a big bonus (first time in years), and the rest had taken off to help the Merry Men, which is why the Merry Men'd decided on that late afternoon. The oil field was as stark empty as a grocery in The Depression. Jim Johnstone was busy cheating on his wife and his good neighbor Holsapple with Holsapple's wife.

That engine block put a nice sized crater in the ground.

"Only a boy named David. Only a rippling brook."

But right near the battlefield motorhome stood another derrick for one of the main wells.

When that engine block put that crater in the ground, a little bit of black oil came up like puss.

Pete got on the radio. "Alright, Remmy," he said. "You need to move it up twenty yards. This is the... no wait. Make it twenty-one. That'll nick the trailer and then slam into the derrick behind it."

"This is the one?" Remmy asked.

"This is the big boy," Pete Taylor said.

Remmy turned and he said, "Alright men, light her up!" Then he breathed in and sang, *"Only a boy named David, but five little stones he took."*

The guy manning the dump truck pulled the sheet off of the big thing in the dump truck. It was that holey meteorite that'd struck down in John David's yard, the one that weighed the exact same as that engine they'd slung. They dumped all three-hundred-and-thirty-three pounds of her right onto the platform and a whole slew of stubborn carpenters started tugging and a tugging on it with the help of tractors and whatnot until they got her in place on them layered tanned cowhides. They filled all them meteorite's holes with shreds of old white work shirts turned black from working in the oil and with scraps of old wife beaters too and a couple of old letters from Grandad Patrick Dempsey. They soaked all that paper and cloth in coal oil and then lit up a handful of shit-filled cigarettes and used *those* to light her up like the world's largest Molotov cocktail, believe it or not Ripley.

"*And one little stone went into the sling and the sling went round and round,*" Remmy sang and then said, "Back, men, back!"

They all backed up.

Remmy finished the stanza. "*And one little stone went into the sling and the sling went round and round.*" Some of the men joined in the chorus, called forth to sing from some antique memory burned into their minds from some event they shared as five year olds. "*And round and round and round and round and round and round and round and round…*"

They pulled the lever and the trebuchet groaned as it whipped that meteorite right out of the resting position across the metal base, a great scratching and grinding sounding quickly as it raked, like the wrecking of a train and many of them covered their ears who were not men that worked with saws.

"*And one little stone went up in the air…*"

The flaming strands of old oily wifebeaters flew high into the air like the streamers do on a phoenix tail, arcing high and far and headed towards that oil field. The math was true — if there's one thing Remmy knew, it was trigonometry.

"*And the giant came tumbling down.*"

He and his men at the fort didn't see it hit, but man they heard it.

According to the men in the demo cars, that meteorite streaked through them grey clouds, lighting them up like lightning and crashed down like thunder. It clipped the top of Jim Johnstone's battlefield office, which buckled in twain and shot forward along with it like a wrapping on some deadly Christmas present. The meteorite with the flaming wife beaters hit that derrick and ruptured the oil derrick, drove it into the ground, ruptured the deeper oil well, and turned the whole middle of that oilfield into something like a bomb. It was flames and fire and the doors of the sea all roaring at them its wrath of having been woke.

"Get to work, boys," Pete Taylor said to the men who was just idling there in their cars all slack jawed. "Come on now, get to work."

Here's the thing. Contractors got a name for being lazy. They get up late sometimes and eat breakfast first sometimes and then wait for the materials sometimes and then wait for the rest of the help to show up sometimes and by then, by God, it's quitting time for lunch sometimes and if you got a Mexican helping you it's siesta time sometimes and then you gotta break for sweet tea if you're Irish and coffee and cocoa if you're not and holy shit it's four o'clock, where's the day gone? Well better get an hour in before we head home. Sure, that happens. It happens because the manual arts have always come before the fine arts: some old hammer swinging fool had to build the Sistine chapel long fore Michelangelo could come in and paint it. That's the way it goes: studs and walls before the painter cometh. So carpenters are just artists with hard hats, don't you see? And even a good sculptor wears a hard hat and flannel and gloves and a dust mask and uses a grinder and chisel. So if they don't get anything done, it's from the messiness of the artistic mind. That don't make them lazy by nature. That makes them untidy by nature. There's a difference.

But you get a damn good foreman to lead them? One they trust, one they'll jump off ten-story scaffolding after? You'd better believe they'll jump when he says jump at other times.

Remmy and Pete both had that kinda foremanship. "Get to work boys," Pete said. "Come on now, get to work."

And boy did they ever get.

They swarmed that old oil field and started out shooting the junkyard cannons, blowing big chunks off of the oil pumps so that some stopped working and others blew up and others turned into giant pillars of fire shooting out of the ground like giant vertical versions of the tailpipes on Bren's old muscle cars. Those hammer swingers smashed into them with those old demo cars. Some of them came running swinging with hatchets and sheet metal swords: one guy had a hand-painted wooden shield made out of the scraps of that five-dollar-a-month Sears and Roebuck kitchen table that could stop a tornado.

A giant spotted boar painted with a blue #3 was running around squealing and hollering its damned head off.

Well by then, the oil barons showed up in their black cars. Oh they were mad. And just like in Bloody Williamson, they'd paid off guards but the guards in this case was the National Guard and the local policemen. The whole damn field's aflame and there's Jim Johnstone standing in the middle of it mad as a hornet-stung mad hatter. Remmy came over the hill with all the boys from the trebuchet and the guys in the demo derby cars just laughing and laughing at Jim Johnstone helpless there like an old fool in the middle of the world's biggest and bestest prank, staring at the giant stone that had fallen on his precious oil field. Some of the boys had brought along smores and weenies they started roasting them over the meteor.

One of them climbed up on that big old bumper of that black car and took a shit.

A policeman got really jittery in the midst of the laughing and carrying on and he pulled off a shot.

The shot hit Pete Taylor in the leg. It hit his femoral artery. Remmy'd read a Hemingway story once that said a severed femoral artery empties itself faster than you would believe. Story was right: Remmy hadn't believed Hemingway until Pete Taylor got shot in the leg and watched him empty quicker than even that old crashed derrick had emptied, that salt water tank at the top. Pete Taylor bled out that quick. And the blood there reminded him of a full bottle of ketchup he'd watched get accidentally shattered while he was flirting some girl at a diner as a teenager, tomato blood slinging everywhere.

Beth had said, "You boys quit it with those bombs. Someone's liable to blow off half a leg and bleed out on the street." Remmy wished then that she'd been wrong. Or he'd wished he knew she'd be right and had quit it with them bombs because he watched Pete Taylor die quicker than flightless birds free falling, blood on the pavement and blood on the streets like someone'd shot an angel and you couldn't recognize its form anymore through the mess and the mangle.

Remmy screamed a blood cry full of Wickannish rage and Grandad Patrick's lust for killing. He pulled out his old longbow and wounded a couple of soldiers and a couple more Texarco bosses. The clean *thwick thwack* of releasing the fletchings and hearing the arrowhead hit true sounded like an old Irish war drum to him, a clicking of crickets or a plague of locusts. He found himself wishing he still had his plane. He could use some air support and chicken bombs. Hadn't a crop duster in a prop plane dropped the first aerial bomb in history during Bloody Williamson?

For a brief second, he had this image in his mind of his daughter Marionette squealing with glee as he'd made the prop plane do a somersault.

"Remember, Remmy," The Good Lord said.

They started shooting machine guns over his head.

A bullet caught the kitchen table shield, but didn't go through.

He didn't have no time to remember. He didn't want and hadn't wanted to stop and make time.

The boys had set down their arms and Remmy'd run out of arrows.

That put an end to everything. Most of them escaped, but the National Guard captured a good many of them and got ahold of Remmy and let him sit while the District Attorney decided just what he'd do.

Half the field had holes in it or mangled steel upon it with the rock that makes men fall right there in the middle, smoldering like the patient wrath of God, slow to anger, rich in love. It was still there years later.

Slow to anger, right in love. At least The Good Lord seemed that way to Remmy at the time.

What followed would make him doubt. It wasn't long before he realized that though jokes and crimes weren't quite the same thing, sometimes jokes have consequences. They were not all the bright image of impunity.

Specially the world's biggest prank.

WILSON REMUS

1978

Judge Penceworth. Remmy thought him a good man who'd gotten him out of some petty trouble in the past and had saved him financial calamity a time or two. Judge Penceworth wasn't bought and paid for, but every attorney and jury member was. So that sucked a big old prickly pear.

Daddy John told him before the trial, "Son, the man who reps himself in a court of law has a fool for a lawyer."

"That or a tyrant for a boss," Remmy said.

"How so?" Daddy John asked.

"You know I can't afford no lawyer," Remmy said.

"Use the one they give you," Daddy John said.

"Oh Daddy, it's nice that you think they're going to give me someone good. Really is. But all them lawyers down at the courthouse are on Texarco's payroll, you know."

"Find one that ain't. They're there."

Daddy John was right. For this one, Remmy wasn't

about to look back years from now and say he should have gotten a attorney. He got the best defense lawyer he could find in Southern Illinois that wasn't connected to Texarco in any way, shape, or relative.

"I need more people that ain't company men," Remmy said.

"Well the judge ain't," Daddy John said.

"Yeah," Remmy said, "but the judge don't call the shots in a case like this. It's the jury. And the jury's done bought and paid for."

"Can't you pick your member?"

"Some," he said. "And we did. Got a couple of my guys on there, but best that's gonna do is keep me from rotting away in a cell forever."

"Well that's good," Daddy John said. "What do you want me to do?"

"Take care of Beth and Bren and Gwen for me while you can. Ain't no way I'm going to let all those boys and their families take the fall for my stupidity."

His Dad said, "I'll come and visit you. I'll take off work."

"And how's mom gonna eat? How's Beth gonna eat?"

Daddy John looked wearied and worn. Like an old workboot. "Shoulda taken more time to play and build and joke around with you, son."

"Yeah, well…"

They didn't say nothing after that.

They started trial lawsuit with the judge. "I'm not gonna lie to any of you folk here," he said to Remmy and the Texarco folk and some of the other defendants behind Remmy who'd gotten pulled up in all of this mess and then to the peanut galley. "This trial is going to take a long time.

I'm a busy man and I'll do my best, but I've never seen anything like this before. Remmy, I know you need a speedy trial and I'm going to do the best that I can, but I've just never seen anything like it so I'm going to do my best."

Remmy looked over at the Texarco table and didn't recognize a single face. He looked behind them to their support side of the room. He hadn't expected to see some of the folk he saw on that side of the room — more of the oil people had been hurt by this than he expected. And many of those oil people were his friends, even *family* — cousins and whatnot of Beth and others. And Jim Johnstone wasn't on that side, he wasn't suing. Over at the Texarco table, a great fat man in a pinstripe suit sat. He walked over to Remmy and said, "I'm the lead attorney for all of Texarco's affairs. I will be the man who sends you to jail for the rest of your life. Remember my face, Remmy." He smiled and put a handcarved pipe in his teeth and lit it. He went back over to his table and sat down. The pipe triggered something in Remmy. He thought of Jim's corncob pipes. He looked over at Jim and saw a defeated, tired man whom nobody had ever loved and nobody had ever welcomed. Jim'd been fired. A man who'd been picked on by Remmy and others for being an oil person just like Beth. And Beth had been right: Jim's family couldn't help it either. They probably came from Oklahoma or some other poor state just like Beth's family cause there wasn't no job there either and the best they could do was take whatever job came along because a man shouldn't let his family starve. So, in a way, Jim had been a victim all along too cause them big companies come in and gobble everything up so that the Jims and Remmys of the world can't do nothing but join in their own demise. And Remmy saw it now with how forlorn Jim looked. He saw the other end of it, how Jim too had built up Bellhammer, Little Egypt right along with Remmy and how they both hadn't had the life they wanted on account of Texarco. He needed to write that man the best apology letter in the world.

Cause his real foe sat at that table. It was right there on the trial lawsuit:

TEXARCO V. BELL HAMMERS

Seeing it printed that aways, Remmy thought he could have made quite the church belfry out of that old derrick. "Go deeper," The Good Lord said. Yeah, and maybe that big old flaming space rock had come from the heavens for a reason: maybe it held bits of the heavens. Maybe he was meant to go to the heavens all along.

The trial lasted a long time indeed. Lots of witnesses. Lots of effects on other cases for some of the other boys, but none of them had the kind of fierce claims Texarco had against Remmy, but that's not how it read in the official papers of the court. In the court filings, it read:

Destruction of property and assault and conspiracy to overthrow government.

...which carried with it a hefty fine and jail time.

It would be a fight. There was no way around it. And he couldn't guess the true depth of the consequences of being the architect of the biggest prank in the world.

They sent him to county jail. County jail's about as far from Camelot as Remmy could get.

WILSON REMUS

1978-1979

THAT FOLLOWING YEAR of waiting was probably the hardest year of Remmy's long life. He once read a T. S. Eliot poem that went *life is very long*. Amen to that. At county, some inmates with lots of money set up a casino. They played blackjack for stamps with one guy acting as the "house" or dealer. He kept the books. Old cards were ripped in half and used as chips. One of Remmy's cellies lost between twenty and hundred stamps every night which was anywhere from three to fifteen dollars and since everything cost twice as much, that's a lot of money in county. At county spades was the main and one of the only pastimes. Spades was played quickly and without fun and with half-attention and without sand bags. Later when he moved up to Joliet, they played it less often and with more fun and sandbags. But not at county. And not at Pontiac neither.

One of the first things he traded for in prison was a haircut. Hair every which way. Another inmate did it.

One guy after another after another like he was shearing sheep. Remmy took a shower. Still hair every which way. He hated hair.

Philippians 4:4-7 came up in church in county. *Rejoice in the Lord always; again I will say, rejoice. Let your reasonableness be known to everyone. The Lord is at hand; do not be anxious about anything, but in everything by prayer and supplication with thanksgiving let your requests be made known to God. And the peace of God, which surpasses all understanding, will guard your hearts and minds in Christ Jesus.* Church wasn't a cathedral. Church wasn't a chapel. Church didn't have no organ. Church was church though cause two or more were gathered. Remmy relied on the verse preached, worked on the thankfulness part and The Good Lord started talking to him more and more, softening his heart. Someone told him he had a light around him.

Remmy said, "That's cause I don't cause trouble in here."

The guy said, "You cause plenty. You got a light around you."

"Well thanks."

"Can I have your breakfast?"

"Sure," Remmy said. "You gonna tell me I have a light every time you want my breakfast?"

"No. Already told you have a light, just thought you should know." The guy smiled.

Remmy nodded back. There was lots of cussing. He thought he got Beth for a moment, all that cussing and no smart thought. He stewed on it.

They liked him anyhow.

See crime like Remmy's made him a villain in Southern Illinois to all that Texarco money, which was most of the people on the outside, but in prison, he was a hero. Especially in 1978.

When he first left from the bailiff's holding, they woke him up and marched him out. He said his good-byes to the boys he'd gotten to know, got back all his clothes. Then he stood in a holding cell forever and a day and proceeded to the worst five hours of his life. They had him shackled hand and foot to this old beat up van with six other guys all strung together like heads of garlic and two shotgun holders at the end, watching them. For Remmy, cause he was so tricky a fella, they gave him one of these rough wood boxes over the top of his shackles to keep him from picking them. It was too big but the marshals didn't care none. The box gave him gouges. Good bruises, you know. All the boys on the truck were talking about sex and guns and heroin all the way to the stop.

Which turned out to be county, but that didn't last long.

Two free letters a week. He got commissary: like script, but prison. He got less phone calls, but they were longer, so that worked out in his favor. He wrote Beth:

I feel safe. Thank God.

Part of that was feeling like a damn hero.

That too wouldn't last long.

He felt like a hero cause everyone in Illinois gossiped, specially the criminals, which they saw as different from snitching. See snitching was telling what a guy did before he was caught. But after he's caught? All's fair game to get mythologized out of all recognition. And in that way, you can guess how a story about a man building a trebuchet out of a derrick and shooting an officer in the leg with a bow and arrow after charging out of an old stone fort will go. It'll sound like the biggest joke in the world. Before long, it'd been a volley of arrows from an entire flank of archers. It'd been men on horseback with pikes made out of old street signs. It'd been that pack of gambling dogs turned wild that he and he alone had tamed and turned loose on the oil man. It'd been men he'd bewitched into bears and packs of wild coyotes — which they pronounced "coyotes" — he tamed and trained along with them dogs. It'd been dragons and steel and dragonsteel, water and fire and firewater, tunnels and flight and the booby traps he'd filled the field with long before he'd landed.

It'd be all the glory and none of the consequences. All bell and no hammer.

They asked him if a certain point was true and he'd always say, "Yeah, but that ain't the half of it."

And they'd go and convene to learn more and more people would bullshit their way into the gossiping crowd and they'd come back, "Did you really smelt your own swords out of the spare bits of old school buses?"

"Yup, but that was the easy part, the foundry we built. There's more."

Even bullshitting like that, Remmy couldn't wait to get the hell out of county. Beth hadn't written back, even as the trial wore on. He couldn't blame her. Such an elaborate prank with such consequences would be a shock even to her and her having lived with Remmy's orneriness for so long. Bren too. And Marionette. Wasn't hardly any letters coming, just his letters going. Some of the Merry Men wrote back but they had to be careful cause even Texarco men was in the jails, opening letters. His attorney came and talked things through. Remmy felt then like he should have just worked less and spent time with them all every week instead of trying to build heaven on Earth. Or maybe his idea of heaven was wrong?

The thing about the nineteen seventies in prison was that it was a time of uprisings. Now Remmy didn't — couldn't — know that, of course, but he just happened to time out right around then. Remmy wrote Beth, *It feels like a monastery or an abbey*, but sometimes people went crazy in that kind of isolation and died. Not everyone can be a monk, you see. Well the seventies was a time for crazy and when you get a bunch of guys down and out, it can get bad.

Started with their work team. They'd been sent out to smash rocks to make room for a road, but those old boys got so damn tired of smashing rocks, that some of them took those same sledgehammers and broke their own legs. They did that till the hospital wing filled up with broken legs. Just legs and legs down the whole wing.[2]

Beth still didn't write back none, though some friends had.

Remmy met a guy in there who'd refused to register for the draft in Vietnam, which Remmy wouldn't have got back when he'd gotten rejected, but he was starting to get more and more how that could be a brave thing to do, even though he once called draft dodgers cowards. This guy was black. The black draft dodgers got five years. The white Jehovah's Witness draft dodgers got two years. Remmy

---

2 Forgot to mention that the bulk of this stuff comes from researching his real prison letters and books and things, but the part you'll recognize the most comes from Zinn's *People's History of the United States*, though I dressed it up a bit to make it feel like a couple or three friends he knew.

hadn't met any five-year sentences for draft dodging so he asked old Hank, "How in the hell'd that happen?"

"I'm black, Remmy," Hank said.

"Oh come on now," Remmy said, "it ain't cause you're black. It's probably cause you were sassing to the judge. How was your hair done up the day you went to court?"

"Afro."

"And what were you wearing?"

"A dashiki."

"See there? Don't you think that made your sentence worse?"

"Of course," Hank said.

"Was that afro and that dashiki worth a year or two more of your life?"

"That's all my life," Hank said, confused. "Man, don't you know, Remmy? That's what it's all about! Am I free to have my style? Am I free to have my hair? Am I free to have my own black skin? Skin and mind ain't the same, Remmy."

Remmy laughed.

"That funny to you?" Hank asked.

"Only cause that's exactly what I told this school bus driver back when I was six. Just exactly. 'Cept the bus was just a flatbed truck."

"I don't got no other and can't get no other skin cause this is me. Take me or leave me and most of them leave me. Leave us all to rot."

Remmy shut up for awhile and then said, "Of course, Hank. Of course. I'm sorry, old buddy."

"You get it?"

"I think I do," Remmy said. "All I've ever wanted was space to tell my jokes around folks I love and to rest with them. Seems when I move towards me, the world tries to make me something else."

"That's it," Hank said, reclining, and closing his eyes. "That's it, old Remmy."

He figured that's why there's a sense among inmates, specially the state kind, that prison's fate. They'll do five years, probably a fifth of it, then a year or so of "papers" (probation), then probably a violation. Then back in jail. Lots of brothers, fathers and sons, husbands and wives, are all in and out all the time. Community can't escape it: cops on the side of the money and money on the side of the largest voting bloc, which is cops and city workers. So prison's just part of life. Remmy begged The Good Lord that he and his family never ended up like that and he begged him to save them too.

All they got — every program and every little thing — the warden used against them later. The right to go to school, go to church, see their visitors, write home, go see a movie together all chained up. None of it was freely given, freely received. Wasn't no rest in that place at all cause all of that was weaponized for correction. All good things became something to lose. None of them programs were *theirs*. The warden treated all good as a privilege to be taken away. That made them inmates insecure and aggravated.

Remmy met a guy who hadn't eaten in the mess hall for four years. Remmy said, "Why?"

Guy said, "Come look." Guy took Remmy to the serving line early and a hundred to two hundred cockroaches — no lie — ran away from them trays. Guy pointed out the grime on the trays. Food that was raw with maggots. "I stick to peanut butter sandwiches, a loaf of bread here or a hunk of bologna there."

"Man, guy, you must have friends or kin."

"That and money for the canteen," guy said.

Remmy realized after trying to write home about some of this that the guards — what his cellies called *housescrews* — tore up letters they didn't like after steaming them open. So he started again, putting less stuff in there about his setup and telling Beth about how he'd felt, rewriting them letters from memory. But she still didn't write back none, so he knew it went two ways for sure, not just cause she didn't initiate any but because she didn't respond to the ones he knew she got.

She really was bitter at him about not knowing about his plan. She was really bitter about how he'd treated oil people his whole life. He shouldn't have shamed her and sent her downstairs. He shouldn't have had fun at Jim's expense all them years.

Around then, he really got serious about playing some poker. Lots of new types. Colorado, Pineapple Twist, Titanic, Cincinnatti, Follow the Bitch, Low on the Flop, 3-5-7. He couldn't wait to teach Beth or Bren or Mary sometime. Well Mary wouldn't never play, but Beth maybe.

A guy there had spent twenty-five to thirty years in prison for allegedly selling fifteen bucks worth of heroin to an informer *that later went back on his testimony*. You talk about bearing false witness, there it is right there. This guy couldn't find a single court — not even the Supreme Court which is supposed to be the last and most sacred hall of justice if the world gives way — to revoke the judgement on this man. His name was Martin. Martin spent the first eight years in regular prison and in the next kind was beaten ten times by guards, then he spent the next three years in solitary captivity, battling and standing up to them powers that be the whole damn way.

You'd have guys who were beat and twisted around them powers's little fingers and then forced to say they're guilty on crimes they hadn't committed — sometimes even to crimes that never happened — say they were entering their plea willingly and not because of promises made and in return they get a less severe penalty. All kinds of guys like that in there among the guilty.

And Remmy was guilty. Hell he was proud of what the state called guilt.

But his attorney wouldn't let him confess on account of all them other boys. And it made sense: they shouldn't take the fall for Remmy. They kept a haggling and haggling with the Texarco attorney and the jury.

Wasn't until later that Remmy realized that he might have done something The Good Lord called guilty, something that made him *feel* guilty. Something that was just wrong.

Remmy saw the worst of it. See that year the Supreme Court ruled that the news stations and newspaper reporters didn't have the right to get into the jails and get into the prisons and talk to people like Remmy. And also that the Warden could forbid inmates to speak to one another, to assemble together, or to spread tracts about forming a union. That's freedom of speech, assembly, the press, and the rest, for the record. A man shouldn't lose what's right to give to all men just cause he stole a loaf of bread and a pat of butter.

Since Christmas, the meanest old prison guards you ever saw started themselves a reign of terror aimed at the prisoners. Four got beaten. One of those four named George ate razor blades and another named Fred swallowed a needle and since the hospital wing was full of busted femurs, they had to rush them boys to the hospital.

Guards got mad about that so they went in and went into the fourth guy that they'd beaten up, guy named Joe, and turned on one of them big old fire extinguishers meant for grease fires, the one with the chemical foam. They filled the room and sprayed it all over him and slammed that steel door in his face shouting, "We'll get that punk."

Well at 9:25 PM, Joe was dead and they told the media it was suicide. Them boys in the cells knew who it was. And Remmy too.

So they went on strike. It was Remmy's idea, but instead of acting like a leader this time, he just kept asking questions. "You boys ever heard of a strike?" He'd tell a story about the way the boys did a strike for beer on the job cause it was hot. "You ever heard of that?"

They started talking about family on the outside that was striking. About all the ways they'd grown up sitting on buses and staying in restaurants and so forth. Remmy told them that wasn't no way of winning a fight and they told him that they'd won a great many things, so why wasn't it? And Remmy couldn't argue with the results, so he tried to learn *their* method. The black men taught him a lot about how to get what you want without being mean.

About two-thousand of them held out in their cells for nineteen days with no food. The housescrews scared and bullied and beat some, but everyone held out. Eventually the warden broke the strike by shackling four of the prisoners naked to the floor of a van. Remmy checked out a book by a Russian guy at the time in which he said, "The degree of civilization in a society can be judged by entering its prisons."

"Ain't that the truth," Remmy said. He'd heard in Norway they fed them like kings and schooled them and almost none of them came back.

Hank said, "The spirit of awareness has grown. The seed's planted."

Remmy realized real quick that the poorer you were, the more likely you were to end up in jail. Rich people didn't *have* to commit a crime to get what they needed. Things like a good doctor or a decent education or a job that pays enough for you not to starve and not to freeze and not to die of thirst. Laws were on the rich people's side. And even if they weren't on rich people's side, hell didn't they have enough money to hire the most vile clever lawyers in the country? And didn't them lawyers weasel their way out and settle for the poor so that the poor would have more money than they could ever count? Remmy wondered how much injustice had continued in the country simply because his fellow poor folk had taken the money instead of a grand jury.

How many of them shoulda got a attorney just like he had?

How many of them coulda afforded it if they had?

Meanwhile, Bren came to visit and I guess he'd got into the dog house with his momma cause he started dating a black girl named "Lucky."

It was not chance that Beth used the phrase "in the

dog house" to describe it: that's how she thought of black people. They weren't people in her mind, they were dogs. Remmy wondered if she ever stopped to think that black folk feel the same way Beth felt when she was called an oil person. That's the dark side of southern hospitality, you see. They love their neighbors really well, they just happen to believe that all the aggravating people aren't really their neighbors. The ones that remember that Jesus himself was brown do real good with the neighbor love, but Beth wasn't one of them. She couldn't stomach it. She couldn't think straight. She was like the old Pharisee who couldn't admit a good Samaritan existed, let alone as her good neighbor. She did everything she could to break those two up thinking that Lucky wouldn't want to marry Bren if she knew Beth would be a brutal mother in law, would she?

Bren told Remmy about it all there during visiting hours and Remmy was so mad, he wrote Beth:

Some of my best friends in here are black guys, some of them smarter and better men than me. Get over that old racist heart of yours, will you? You want grandbabies bad enough, ask yourself this:

If you could have a black grandbaby tomorrow or a white grandbaby in ten years, which would you pick?

He knew from Bren that Beth had been giving Marionette trouble about having a grandbaby, begging her to go get pregnant, to get cleansed by the priest or to get some sort of miracle fertility drug or something. All during the damn trial! Guess folk need more babies in hard times than easy. Beth tried feeding them aphrodisiacs first: oysters, chili peppers, avocados, chocolate, bananas, honey, coffee, watermelon, pine nuts, arugula, olive oil, figs, strawberries fresh from the garden, artichokes, pomegranates, cherries, pumpkin seeds, and whipped cream. The whipped cream didn't have nothing in it, she just thought playing in it with her fingers in front of her daughter and son-in-law would get their fancy to going about parts of the body they could cover with it.

Add that to the constant begging for them to give her a grandchild and you'll see how tense a dinner could be, mother to daughter and a young man besides.

What Beth didn't know is that Mary and Dean didn't need no help making love. They made love plenty. They'd already been using whipped cream long before she played in it. Sure the aphrodisiacs made it more likely to happen, but Mary was trying all she could. She asked Dean for Rose Quartz jewelry. She danced a baby dance naked in front of Dean during supper one night when it was a full moon. She wore a pouch of rosemary around her neck when she was making love and sometimes wove it into a little headband. There was an old Scottish rhyme that went, "If you rock a cradle empty, you'll have babies plenty." So she'd stop by Sears and Roebuck and rock all them cradles once a week. She drank from the blessed well in the back of their new house, the one that didn't have a lick of salt in it or a salt lick near it. Mary had a little hen who laid a bunch of eggs so she grabbed up that thing and tied it up to the bed, which could have worked, who knows? They never got to find out because once the bed got to rocking, the hen got scared and even though chickens can't fly for shit, sometimes they can flap hard enough to get up in the air a few feet and that rope was long enough that it whipped that chicken around so it flew up and arched like a rockheavy sling and landed on Dean's naked back and started tearing at him before they got very far in the baby making process. So he had big old scratches and gouges and was down for a few weeks with no drive to make love or make babies or ask Mary to take her clothes off.

Mary cried around then and started making spears, whittling like a boy scout in the backyard.

But Beth didn't stop. She went to feeding her daughter red raspberry leaf, nettle leaf, dandelion, red clover, and alfalfa. The alfalfa she put in a rhubarb pie. Dean thought the alfalfa tasted like horse shit.

So when Beth got the note from Remmy that said, *If you could have a black grandbaby tomorrow or a white grandbaby in ten years, which would you pick?*

She really, *really* had to think about it. But she didn't write back.

Meanwhile, Remmy had wrote Bren that there was... *lots of nicknames in this culture. Slim, Mikemike, Scarface, Belize, Bones, Jimbo, OB, O, Jay, Fab 4, Lil G, Bib Baby, Lalo, No Love, Too Tall, Two Six, E, Double R... All night guys are talking so loud about the "ho's they be pimpin'" + "the bitches they got bringing dem money." Can't sleep. Being in jail is called being "down." I suspect this comes from that old saying "sent down the river." Also groups who've got each other's backs are called "cars," probably because they ride together.*

Bren read all that to Beth and Mary.

All Beth said was, "Those nicknames... those are Lucky's people I can imagine. See what I mean, Bren? We don't want our boy dating *that*."

Bren went back to county and said, "Momma done scared off Lucky good and sure, Dad."

"Why? How?"

"Called her names and words and things I never heard momma use. Cursing but Lucky wasn't no curse to me, Dad."

"Well you can get her back," Remmy said. "Some girls just get feelings hurt is all. Just treat her kind and take good care of her."

"No, dad, Lucky went and married my best friend."

"Ah."

They sat and listened to the other prisoners talk about babies and knitting projects.

Remmy said, "Well go on and date whatever color girl you want and I'll deal with the fruit if your momma ever answers my letters or comes to visit or when I see her in heaven."

"In heaven?"

"I figure she'll be keeping The Feast with every tribe and tongue."

"Like a potluck. You know it's nice to share with other poor folk, dad."

"That's the thing. Every rural white Republican, urban black Democrat, and Mexican immigrant has that in common. They all think they're superior to each other, to women, and to homosexuals. They all drop out of high school. And they're all poor as shit and need each other's help to make it. The whole race barrier is just a language barrier. Money and power's stepladder divides way more than race."

"People sure cuss a lot," Bren said.

"You know, Buddy, I never cared much about school."

"Me neither, dad. Waste of youth."

"Yeah but it has a value nobody talks about. Everyone in county with high school or better's quiet and mannerly. They can talk about how they feel down deep or talk through a problem or tell a story about how hard it is. But without high school — which has got to be nine out of ten of these guys — it's the other way around. That's where all the vulgarity and volume comes from, the zoo. They can't talk about what hurts. They can't listen. They talk over each other all day long like their brains only handle a few seconds of messages coming in before getting clogged with messages going out. So 'old man' becomes 'fuckin old ass mother fuck old man. Haha! Yeah! Ah! Yeah! Old, family! Old!"

"Reminds me of Grandma Broganer."

"My mother?"

"Yeah."

"No comment."

"Grandma wouldn't cuss," Bren said.

Remmy said, "She was just like your mom in that."

Bren said, "But Grandma asked you if you *had* to cuss."

Remmy said, "Like Beth. Wasn't educated neither."

Bren said, "Only way she knew how to argue was shouting at your or making you feel stupid or like more of a sinner than she."

Remmy said, "Well being uneducated… look she could only get a few seconds of incoming messages before her mind got filled up with outgoing."

"Bad listener too. Talked *at* you and not *with* you."

Remmy said, "Didn't know how to think and write and speak for herself or listen to folk on the other side."

"That's why she asked you not to cuss: she's doing the same thing with the letter of the law and her religion and rightness as these guys in here cuss. She's breaking the third commandment too, woman like Grandma who's religious in all the wrong ways, using God's name in vain."

Bren laughed.

"How's the business?"

"Couple teams under me. We took on steel studs."

"I's always scared of them."

"Got us a lot of corporate work."

"Company man."

"Sorry," Bren said.

"No, no. You gotta do what you gotta do sometimes. There's dreams and there's realities and best we can do sometimes is push the real further up and into the dream. I get that now."

"Thanks dad."

"What companies?"

"Building restaurants and department stores. Built a big old warehouse for Fabic tractor."

"One I turned down?"

"Yup."

"Proud of you, son. You're braver'n me."

Beth still worked the books for Bell Hammer Construction Company under Bren, paying herself both her share and Remmy's, which was nice because Beth wouldn't have been able to keep the house no ways otherwise. Between that and the IN CASE OF BLOODY WILLIAMSON money sack, she did more than alright, dividend paying stocks and whatnot. And even then, Bren had plenty of money to buy many cars and as many lotto tickets as he liked, so helping his parents didn't hurt him none. As one of two kids, he'd get a good chunk of it back some day anyways.

She refused to write Remmy back or visit. Still mad, let me tell you.

Beth invited the ladies over for cards as much as she could, specially the widow Taylor.

The night she refused to write him she had the girls over for cards. The widow Taylor and Mrs. Johnstone and Mrs. Holsapple and Gwen and a few others. Beth tried and tried to get Marionette to play cards with her like they used to when Mary was young but between the hassling over her getting pregnant and Marionette hating how her mother gossiped, Mary didn't want none of that. The girls came over and Beth had baked bread and cooked them steaks and had mint juleps ready.

"Mint juleps," Gwen said.

"What?" Beth asked.

"I just don't know what to do with these fancy big city meals and fancy big city drinks." She held up her mint julep to the light and stared at the bourbon and leaves.

"Well it ain't that hard to make," Beth said. She tucked her chin, thinking Gwen was giving her a praise.

"How fancy you've gotten without Remmy here," Gwen said.

"How you mean?" Beth asked.

Mrs. Holsapple said, "I don't remember you making fancy meals like this for Remmy."

"Remmy didn't like mint juleps. I do," Beth said.

"So now he's in county waiting his sentence, you just going to do what you want?" Gwen asked. "You just gonna act like my brother's dead and gone? I don't think Remmy allowed liquor in his house."

"*His* house," Beth said. "*His* house? *His* house is that eight-by-six box four hundred miles north of here. This is *my* house now, sister. *I'm* the woman of the house. And right now I'm also the man of the house cause ain't no man left to play house with *me*. And the child, too, come to think of it, since Mary and Bren are gone and moved out of the house to St. Louie and wherever else." She sipped her drink. "And I say we're having mint juleps. Now sit down and shut up and deal your hand, Gwen."

Gwen sat. A long and loud fart sounded out.

The women all blushed and giggled.

Gwen pulled out a whoopee cushion. "Really, Beth? I didn't know you had it in you."

Beth sassed, "Just cause I like mint juleps don't mean I don't want Remmy here in spirit."

"It's a whoopee cushion," Gwen said.

"Best I could do," Beth said. "I don't know how he does it. Best joke in the world. Your deal, sister."

WILSON REMUS

1980

A GUY IN THERE, in county, got a few different colored candles and melted them in the microwave and poured them into different little boxes. Then he rolled up pieces of paper real tight and cut them so that they frayed. They looked like small paint brushes. And that's how he used them. Wax and water heated make the paint. He did commissions. Did a Water Lilies. Did a Mona Lisa. Did a wax waxing moon, which only he and Remmy found funny.

That painting wax sounded nice and clever to Beth. She'd been painting a lot herself. Painting lighthouses left and right. That's the only thing on her mind: just canvases and canvases of lighthouses in the dark and the fog, sometimes in the day with nothing to throw light on. The lighthouses in her mind and the note about the wax paintings finally did it for her.

Beth finally wrote him back:

Porch is busted.

She hadn't told him that the stories he'd been telling her haunted her dreams. That she'd been praying every night and every morning for his safety. That that house sure was quiet without him. That she was excited about him. Remmy knew all that, but it took him awhile to get it out of her in a letter. He wrote back:

I'll fix it soon. Sorry the house is so quiet. Throw a handful of sawdust up in the air on Fridays, spill some motor oil on Mondays, and run the tablesaw for an hour on Wednesdays and it'll help. Maybe read one of your books and imagine it's me.

She visited after that in the family meeting place. "Thanks for the drawing. It's nice to see where you are so I can think about your days."

"Thanks for coming love dovie honey bunches sweetie pie sugar lumpkin."

"Lumpkin's such a gross word, you know I hate being called that."

Remmy said, "God I love you, hippo shit."

Beth said, "God I love you, felon. Sawdust and oil helped but you'll *never* talk me into running a table saw without you. Less noise is nice."

"I'll send you one of my own paintings." Turned out to be a full sized canvas painting of a lighthouse. Good Lord! He'd never seen something so beautiful in all his life. He wished he had a frame. Maybe he could build a frame. Yeah, he'd build a frame.

Frames were hard, because any metal or wood that wasn't nailed down or chained up to the wall got picked up by the housescrews and thrown away. Some people had shanks and things, but Remmy didn't want to get caught with no shank, so that was out. What he did was take strips of old shirts and sheets and corded them together into a braid that he could wrap around the edge of the thing like sailors will do a painting or a table. A rope frame. He got an idea.

He started saving bits of shirts and sheets and towels and any old thing he could find and he started making him a rope. Thirty feet. Forty feet. Fifty. He wove and braided and wound that thing until he could barely hide it in his pillowcase. He wound it up and then late one night he tied it to the cell bars and threw it out over cell block 9 and down that long drop to the floor of cell block 1. They share all the same airspace along that wall in Joliet, see. And he waited and waited, listening to his cellie sleep, giggling to himself until the buttcrack of dawn he shouted, "HE'S MAKING A RUN FOR IT! PRISONER OVERBOARD! HE'S A RIGHT WINNING ESCAPE ARTIST HE IS!"

The whole cell block woke up and then all the floors below it and the cells that saw the rope dangling down in front of the bannister in front of their cells went to hooping and they all got excited cause someone had escaped. The housescrews came up to him and said, "Who escaped Remmy? Who was it?"

"Why it was old Jack," he said. "You know."

They went on high alert and turned the guns off safety in the towers and rang the bell and sounded the air raid horns just like Pearl Harbor and men was ringing their toothbrushes and metal cups along the bars of their cells, making a godawful racket.

Hell it was noon before they came back and said,

"Remmy, there's no Jack on this floor. Or this wing for that matter."

"Musta been the ghost of Jails Past," Remmy said.

And all the prisoners cackled and hollered and Remmy got a night of solitary for the prank, but the warden didn't mind that much since it worked out kind of like a training exercise.

That night was crazy. Guys called a Bible study. It was nice. In the middle of it, one new guy showed up and after he found his bed, joined the group. Then he said he had just been in OKC and studying a lot with a group in that jail.

When he moved prisons, he was bummed to leave all that and prayed to God that he might find a Bible study somewhere, somehow there in Joliet. Two-fifty inmates on his flight to Saint Louie. Ten of them continued on to Kaskaskia. Two ended up in their wing and right exactly during that there Bible study. Pretty neat. Remmy felt his similar prayers were answered by coming to county. He put his hope in the Good Lord for whatever stop was next.

Something told him that Pete Taylor would have enjoyed a group of felons having church together in prison. Some of them looking at twenty years, some of them looking at six with young children at home thousands of miles away, no hope of seeing them for the entire sentence. Another guy there did a Bible by mail course from the American Bible Academy through American Rehabilitation Ministry out of Joplin, Missouri. Joplin, Missouri. Remmy never thought he'd hear the name of that town again. He could have been building baptistries for prisons instead of death towers for kings. Not much difference between the two, just depends on if a man's laying down his life or taking someone else's. Whole thing sounded like a joke to him.

Pete Taylor also probably would have enjoyed that secret pair of nail clippers dangling from under one of them tables, dangling by magnets. You sit in the seat, search for them without looking on account of the guards, and then you cut your nails under the table out from their gazes.

Man oh man he missed Grandad Patrick. And Beth. And Pete Taylor.

They brought them chicory coffee in the mornings. Smelled to Remmy like one of the first stages in Beth's process of baking a chocolate cake but tasted like tea steeped in beef stock or maybe percolator coffee brewed with the water from that pan beneath a Christmas tree. Man he missed divinity. And black licorice.

Guards would do anything for the right price in county. Hector, one of his cellies, used to talk about paying one thousand dollars to have sex with a guard three times through the bars of his cell. He said, "She looked like a horse. The ugliest woman I've ever seen, but I *had* to do it. I just couldn't take no sex no more." Hector was not endowed like many men in the Broganer clan and therefore Hector reaching through the bars didn't seem like a stretch to Remmy.

Literally.

Another guy said a guard helped him escape once. With the uselessness Remmy saw just after he threw a rope out the bars, that didn't seem so far fetched to him. And he's the one labeled as a criminal and society's deviant. A guard helping someone escape though...

He couldn't take his friends getting in jail too. Jail was hell. He owed it to the Merry Men to keep them out.

He owed it to Camelot to lay himself down so it could live on. The prank too, even. He knew that Grandad Patrick had gotten off scot free killing coal scabs with his bare hands all them years before and had hid away in Boulder and the rest, which was damn near underwater now like the judgment of Noah's time. Remmy couldn't run like that, not when he knew he was guilty. Not when he could get his friends off for their serious lack of guilt. This wasn't Bloody Williamson. They'd made their point and Remmy was the only one that'd done wrong on his side, the way he saw it.

Besides, he owed it to Maid Marrian to take a bullet that she might live to have a grandbaby. How could he keep her and them out?

Remmy had an idea.

So his attorney got together with the Texarco attorney cause they were at a standstill in court. Texarco attorney's name was Deemer. Remmy's attorney was Old Slapjack. They all sat down at one of the metal tables in the pen.

"You need to confess," Deemer said. "It's not looking good for you and we're ready to wrap this up."

"Can you reduce my sentence to ten years?" Remmy asked. "Ten years is a nice round time. Man can learn a lot in ten years."

Old Slapjack said, "Only way my client's fessing up is if you sign a contract saying you won't press charges against the others."

"Why would we do that?"

Old Slapjack pulled a copy of the report Remmy'd stolen from Jim Johnstone and slapped it on the table.

"Cause otherwise he's going to blow the lid off this coke can."

The Texarco attorney Deemer stared at the cover page of their company's own scientific survey for the better part of five minutes: *Arctic Projections of Worldwide Warming and their Effects on Arctic Drilling*. Remmy shifted awkwardly — he knew how to haggle over the price of bargain bins and barbed wire, yet this offset him. He didn't know what else to do but stay quiet because Old Slapjack did nothing but stare down the other guy with his good eye like he's sighting in a rifle.

Deemer breathed in and said, "Okay."

Remmy laughed.

"But Remmy serves the maximum the judge will allow."

Remmy stopped laughing.

Old Slapjack said, "Whatever the judge allows. Long as you get the other guys off."

"Deal," Deemer said.

There were papers. John Handcocks.

After Remmy said, "What was that?"

"You got what you wanted."

"I didn't want to go to jail."

"Then next time don't do something so dumb."

In the court room, Deemer dismissed charge after charge of guys standing in the peanut gallery and sitting on the benches and the rest, all on the signed condition that Remmy pled guilty.

Then Old Slapjack counter-sued Texarco with a whole set of new charges.

But then all those Merry Men who'd had charges dropped stood to defend Little Egypt. Old Slapjack took witness after witness, one right after another. They talked of poisoned wells and stolen pensions, of families relocated in order to keep people from assembling and strike breaking and overdue union fees. They talked about newspapers bought out so that they couldn't talk about it no more. They talked about public parks bought out so they couldn't meet in public no more to talk about the issues amongst one another. They talked about how Texarco'd set up an oil-worshipping religion and made it really hard to petition the local rule outside the state. They talked about how the company forced people to let their bosses stay when they came in from out of town and how they'd bought out the police for what they called "civil forfeiture" where they could take any cash or papers or property during what they called "routine inspections." How sometimes they'd put people in jail for the same thing twice in two different counties and even recorded them so that they testified against themselves. How they had held secret trials that dragged on forever in order to keep unpersuaded people in jail and the threat of doing that to shutting up about the things Texarco was doing. How they'd bought out judges and did away with juries and made it so that people they put away had bail put so high that there was no other choice than some of these poor workers getting hung out to dry in prison for a good long while. This and other things like oil leaks that killed off a ton of wildlife in the national parks and so forth came forward.

They never mentioned how they were melting the whole damn planet.

They didn't need to.

Judge Pennyworth nodded his head.

The reaction from the jury was mixed.

Remmy took the stand.

"Remmy, did you destroy a piece of property?" the attorney asked him.

Remmy looked right at Jim Johnstone. "Yes. Many."

"How?"

"I made a trebuchet—"

"A what?"

"Big ass old catapult. Made it out of an oil derrick that fell in my yard all them years ago and its saltwater ruined and contaminated and adulterated and befouled all our wells. Guess you can say I fight Texarco with Texarco."

Carpenters nodded.

"Did you make cannons for an armed rebellion?"

"I made cannons for blowing up their pumps. That's about it."

"And the National Guard?"

"I don't care about nothing but scaring Texarco outta Southern Illinois."

"So you admit to this violence."

"Oh yeah, I'm guilty of that. Guilty of that because they're guilty of so much more."

It went on like that, but there wasn't much else to it in terms of the law. The sentence came to the judge and he read off Remmy's sentence: ten years in prison, no parole. Remmy recalled how his Uncle Charlie'd been a bastard and had once murdered a man for flirting with his wife, double barreled shotgun to the belly. He'd gotten ten years too. Nice round number, I guess. Charlie's brother — Remmy's other uncle — was a Pentecostal minister that'd healed some people. Remmy figured he was halfway between the two.

Judge Pennyworth after he read the sentence kept reading.

Or faking like he was reading.

Remmy thought he might be ad-libbing, talking out his ass like all the other men will do.

"Jim Johnstone, one month in jail."

"WHAT?" Jim cried out.

"Order or I'll hold you in contempt. Texarco to pay a fine to the people of Marion and Clinton counties of ten million dollars to be turned into a sovereign wealth fund, invested in safe ventures, and to pay a dividend in perpetuity to the citizens of Little Egypt."

One of the Texarco men turned to Jim and said in front of everyone, "You're fired."

"WHAT? Don't you think about firing me! I've worked for you for… for twenty-six years! Since I was seventeen years old! My daddy gave his whole life to you."

"Order."

"You're fired."

"No, don't you say that. I can make it better, I can—"

"Order!"

"You're fired, Jim. Be quiet now."

"I SAID ORDER!" Judge Pennyworth said and blasted his hammer hard as Thor. But no bell rang out.

The man Jim Johnstone looked really old and wept silently with his head down. His wife looked like a corpse. His children looked around awkwardly, unsure what to do in this situation.

Everyone filled out, talking about what had happened: how strange it had been. How exciting. How dreadful. On his way out the judge said, "Sorry I couldn't do more than that, Remmy. It was pushing the limits as it was, but I couldn't let no injustice be done." He patted Remmy's shoulder.

Remmy swallowed and said, "Just make sure I can write my letters."

"I will, Remmy, I sure am sorry."

"What about the other guys?" Remmy asked.

"Texarco felt bad enough sending you and killing Pete Taylor, I think. I don't know why else they would have signed that deal of yours. No one else's going to jail."

"Like Bloody Williamson."

The judge nodded and smiled. "But not quite."

"No," Remmy said. "No not quite. Different in the right way."

"Most of the right ways, anyways." After it got quiet, Remmy's sitting there in jail waiting to be moved. "Good Lord," he said in his cell, "Where on earth did I go wrong?"

"I think you know," the Good Lord said.

"It was that groundhog I wounded."

"Before that, Remmy. You never went deeper like I told you to go."

"I've been chasing the wrong thing?" he asked the Good Lord.

"You've been chasing a wrong many things, son."

They moved him to Pontiac.

The idea of a strike took deeper root in Pontiac than county and they suffered from hunger strikes and beatings. Refusing to work the kitchens wasn't much suffering, but it fully shut down the joint. All sorts of nonviolent protests. After spending fourteen to sixteen hours a day in their cells, mail read, reading material burned, visits from family through a screen, base medical care, harsh parole and the average parley on their hearings lasting 5.9 seconds — and that's a fact on the public record, racism every which way, that's the only option they had. Martin Luther King had said that a riot is the language of the unheard. The best learn nonviolence because they don't stoop to no mean pranks and surely to no war. And so, when they got rid of the food and went on strike, the superintendent at Pontiac asked, "Why are they wrecking their home?"

Dumbass.

Remmy was crying because that prison yard was the first place he'd ever seen in Southern Illinois without any racism. Hard to believe now, since prisons are so rampant with racism, but that's the way it was that year. Hank, the black guy who'd followed Remmy to Pontiac, said, "I never thought whites could really get it on, but I can't tell you what the yard was like." Hank cried it was so close, everyone so together.

It was so close that the white gang and the black gang had kept up the air of racism to the guards but behind closed doors they were joining forces. And they'd worked out a signal.

The main signal starter was the guy that had told Remmy he had a light about him. He came up to Remmy and told him the same thing and asked for Remmy's breakfast and Remmy offered his breakfast over, laughing cause he knew that's all that guy wanted. Well the moment Remmy passed his tray, hundreds of guys in the mess hall upset their trays and then laid down on the ground, flat. It spread like dominoes to the guys in their rooms and the guys in the yard and everyone — two thousand of them — all laying down in protest.

Remmy was the last man standing so he shrugged, even though he didn't know about the signal, and sat down with them all.

Five days of no food.

Five days of no work.

Five days of no sleep.

Five days of two thousand prisoners not budging.

What normal suburbanite would've been strong enough?

Not Remmy before prison. He'd taken the easy way out before: consequences last, meanness first.

So the Governor stamped down his seal for a military attack there on the prison. He pulled in the national guard. He pulled in the prison guards. He even pulled in the local police. And all of those fools went in there loaded up to the brim with them automatic rifles and them carbine rifles and them submachine guns like the damn mafia. They went on a no-joke, wartime press on the prisoners. None of them had no firearms. They weren't even holding brooms or nothing — they was just sitting down in the yard at that point, refusing to move, locking arms like the suffragettes. All them big city papers that'd lied about Bloody Williamson and unions told everybody that nine guards were held hostage and that nine guards got their throats slashed by prisoners during the attack. Murdered while hostage. The autopsies proved all that to be bullhocky and codswallop: the nine guards died in the same hail of bullets that killed thirty-nine prisoners in the massacre. They'd shot their own.

Warden decided to spread out those who had been

known to start revolutions, whether they'd been the brains behind the protests or not.

So they shipped Remmy off to Joliet.

That year before and this one was the years they filmed *The Blues Brothers* just right outside. It was pretty weird watching that show in later days with his grandkids, knowing he'd been in prison while it was being shot.

Commissary they ordered over the phone and was delivered the next day like a pizza. He could hear the TV in the main room.

He found out at Joliet that it and many other prisons they'd split up based on crimes. So they had a murder block and a violent drug block and a robber block and a gun running block and the rest. He was in the armed rebellion and civil resistance block, which took a major load off. The Joliet boys in that rebellion and civil resistance block were the nicest guys he'd met in prison and not a one of them had a reason or proclivity to judge him and kill him. Not that they would *him*, as such, he was just relieved to know that even if he wasn't an accidental hero of The Resistance, even if he had been the sort of guy everyone wanted to kill like the ones they called "chesters," he still liked to know that they weren't the killing type. Elsewhere snitches get stitches. Elsewhere the chesters got mop handles shoved up the butt until they died. None of them cared about the facts, they just labeled you and attacked you, which was ironic coming from "murderers," "drunks," and "thieves." In Joliet, he was safe for sure.

Turns out prison clothes are made by Bob Barker's company.

Took five years out of prison for Remmy to realize there

was no relation the gameshow guy who made sure your pets were spayed or neutered.

Which stunk because Remmy had some good jokes about making sure to spay and neuter your prisoners.

Beth came north for a visit. They had a separate room for them — helps sometimes to be decent to the guards. Guards checking on them every so often, but mostly left them alone to talk. "Bren met somebody."

"Really?"

"He keeps talking bout this girl he likes a heck of a lot. No kids from Mary. I want grandbabies so bad I could carry them myself. Maybe we should have another."

"Your body can't take another baby, Baby. They'll come you just got to be patient."

"Coming from a prisoner, that's saying something," she said. "The old redbud fell down in the backyard after a lightning strike. Burned it and those bloodied heart shaped leaves in a bonfire in back and invited the neighbor kids over. Johnstone came over and Jim wouldn't leave me alone. I told his wife to tell her husband to leave me alone and he minded then, blushing."

Remmy said, "Tell Jim I'm sorry for what I done, but tell him not to come near you again."

"I think he's snapped, Remmy. I think he's come unhinged now, fired and all."

"You watch him. We don't need him having a come undone. Beth?"

"Yes my sweet felon."

"Good Lord I screwed up Beth and I am so, so sorry I let you down like this."

She weighed it until their time was almost up and as they came to take her away, she said, "I forgive you."

"Ten minutes," the guard said and walked away, leaving the door to close.

They hugged tight. She held up her man who nearly collapsed in her arms — her man who had tried to hold up a whole county.

He held her face. "Man can do a lot in ten minutes with a closed door," he said.

"Oh Remmy, quit, they'll catch us."

"And do what? Throw me in prison?"

She grinned the very impish grin that belonged on *his* face and hiked her skirt before straddling his lap…

Two nights later, Beth's tenderizing a chicken with that big old meat tenderizer and heard something behind her and it was Jim come to say sorry like a sales pitch and bringing over snicker doodles.

Beth told him Remmy was sorry and to leave her alone. And she said, "I don't know what you're doing here, Jim, but you need to get on."

And he said, "I just can't live anymore without being a company man. I don't know what I might do without another oil person in my life."

"Go on home, Jim," she said.

"And it's all Remmy's fault."

"He said sorry. Now get."

"Sorry? Sorry for taking away the only thing life ever gave me? Apolo— we're destitute, Beth. We're poor as shit now, don't you see? You leave the race alone and we just return to white trash cause them WASPs don't want to help us, now do they? I have no one. I have no friends. I have no neighbors. No secretary. I have *no one*. I can't just move back to Oklahoma. My wife won't touch me — hasn't in years. My kids won't mind. What can I do?" He looked at her.

"Jim, please."

"Remmy has it *all*."

"Remmy's in *prison*," she said.

"He still has you."

"Yes he does," she said.

"I just can't go another day without you, Beth. I'm crazy for you."

And she said, "You're talking crazy. I'm a married woman and you're a married man and I'm in love with my husband and don't want nothing to do with you."

"Nobody does." He spat. "You're practically a widow," he said. "You need someone to comfort you," he said.

"I'm quite fine with my friends."

"Oh now, don't act like that," he said and he moved slow towards her. Slinking. "We could share friends."

And she said, "Don't come no closer or I'll tenderize you." She raised the hammer like the hammer of Thor and Penceworth.

And he said, "I need tenderizing. I'm too hard of a man, you see. I'm too lonely. I've lost it all." And then he came at her.

She swung at him, swung for the fence.

And he caught the bell hammer of her kitchen and yanked it away and he pushed her up against the wall.

She tried hitting him and hitting him and he didn't pay it no mind. So instead of screaming and hollering and punching, she leaned right into his ear so that her lips would tickle him when she whispered instead, "Wait two seconds and listen. Two seconds."

He stopped. He looked up at her.

She leaned close and whispered again in his ear. "I know you Jim Johnstone. I do not hate you. I do not love you. But I understand an old oil person when I see one. Those eyes of a little scared boy. A scared little oil person coming in to tell everyone that a fire has started in the south Pacific. A scared little oil boy whose daddy forced him to move away from their farm in Oklahoma just like mine did to me to work for a nasty old company that steals from poor folk just like my daddy'd done to me and Daddy John'd done to Remmy. A scared little oil boy whose notebook my husband pissed all over. I don't hate you. I don't love you. I understand you." Then she did the only thing she knew how to do: worship and thank the Good Lord no matter the circumstance. She sang in her whispering alto as she held him.

When peace like a river attendeth my way
When sorrows like sea billows roll
Whatever my lot Thou has taught me to say
It is well, it is well with my soul
It is well
With my soul
It is well, it is well with my soul
Though Satan should buffet though trials should come
Let this blest assurance control
That Christ has regarded my helpless estate

And hath shed His own blood for my soul

And she sang the chorus as he grunted trying to hold back tears before he buried his chin into the nook of her little collarbone and made her blouse wet with water salty enough to rust chrome right off a bumper.

My sin oh the bliss of this glorious thought!
My sin not in part but the whole
Is nailed to the cross and I bear it no more
Praise the Lord, praise the Lord, O my soul

She sang the chorus staring in his eyes and there was a horror there. A dread of something.

For me, be it Christ, be it Christ hence to live:
If Jordan above me shall roll
No pang shall be mine, for as death as in life,
Thou wilt whisper thy peace to my soul.

She refrained again to that old sinner. And his eyes turned into the eyes of a young man just moved into town and there in the one room schoolhouse.

But Lord, tis for Thee, for thy coming we wait,
The sky, not the grave, is our goal;
Oh, trump of the angel! Oh, voice of the Lord,
Blessed hope, blessed rest of my soul

He crouched over the corner of the kitchen covering his face as she sang the chorus again. He started crying and then he got up as she started to sing the last verse and ran out of the house.

And Lord, haste the day when my faith shall be sight

The clouds be rolled back as a scroll;
The trump shall resound, and the Lord shall descend
Even so, it is well with my soul.

"Come back, Jim," she said to no one in particular. "Let me hold you awhile longer. Let me hold you while you cry."

She wrote a letter to Remmy about the whole ordeal but never sent it.

WILSON REMUS

1981-1985

Jim Johnstone made a last supper for his family and pulled up some of that hemlock he grew in his backyard and steeped it with the fresh mint to make a tea. He went to feed that to his wife and kids and grandkids, but then he lost the stomach for it and poured it all down the drain. None of them were any the wiser. None but his wife, who had figured out what he was like with other women. She'd watched Jim brew that tea and brewed one herself, died as sure as Socrates, leaving her boys without a mother.

That was several months into the year, you see, nine months after Beth'd visited Remmy for about ten minutes, and it was a week of torrents and fall rain the week that Beth went into labor. She wrote to Remmy that it was a blessing to be a mother again, a grandmother on the way, that only The Good Lord could bring so much good out of so much bad, something like us storyweavers, something like clockwork that ticks one tock towards redemption even from so low a point as six-o-clock. That only The Good Lord could write a good ending from the worst kind of things. It poured and poured and she worked in the garden that week. She never did send that letter she wrote. The letter that started with the words: *I'm pregnant.*

They took her into the hospital. She wouldn't be having no baby in no field. They took her in and she had a horrible cough from the storms. Pneumonia. "This is a blessing," she told the doctor through that coughing.

"Not in the rain and the wet," he whispered to the nurse.

Bren realized too late that Remmy hadn't heard about none of it. Bren found that letter that started *I'm pregnant,* and he sent it along with one of his own to tell all three bits of bad news to Remmy at once: the almost-rape, the hospital visit, and Bren's own struggles with the business and dating.

In reading it all, Remmy realized that after all that fight in her, the song — the song of the woman with the kitchen's bell hammer — had defended her like St. Patrick's breastplate, defended by a song. He wept at that. While crying, Remmy recalled sitting in the Kaskaskia River valley that became Carlyle Lake when his Daddy John'd said, "Drink it up, gang."

He hadn't listened. He hadn't drunk it up.

That had been the first time Remmy'd seen his father cry since the incident with the milk wagon and he cried there in his cell thinking about it. And it felt then in his cell like those waters had come to wash him clean away.

Remmy prayed for a verse and he flipped and dipped. The Bible opened to Paul and Silas in the prison yard and their song of freedom rising to the stars, like that song says. And he sang *it is well* just like she had, echoing her soul. A song for the defendant. And he sang, "You can take my wife away, I'm gonna let it shine." And some of the prisoners joined in singing about how they could take their life, take their land, take their mattress, their food — none of that would rob them of the light of life and the sound of a soul secure in some coming bodily resurrection. Not resuscitation. Not reanimation. Not some undead army. Resurrection. For if the Good Lord could make it once, he could do it all over again.

"I had a son," Remmy said. "His name was Toby."

"You have a living son," The Good Lord said. "Two of them."

"I have another son?"

Bren came then and talked about the nameless child that had been born of their ten-minute union the year after Jim Johnstone had threatened his wife Beth. Remmy wanted to kill Jim and that's most of what he thought before he realized he really did have another boy.

Bren said, "I don't think you've thought it through: a baby given to you to love just like both of them oil people in your life. If Jim'd been named Beth, you'd have loved him forever."

"A baby and two oil people in the hands of a sinner like me," Remmy said. Not like oil. Beth had made it better, made him rethink the whole oil people thing and put the blame where it needed to be: not on Jim, but on Texarco. This child was a spring of living water, the geyser to flush out the oil well's blowout. *I want a well*, he thought.

"I will give you Water," The Good Lord said again.

Okay, then, Remmy thought again. To Bren he said, "You name my child Water."

"Water?" Bren asked.

"That's right, not Oil. Water. Water Jim. Water Jimmy like Wilson Remmy."

"Water *Jim*?"

"After Jim Johnstone. This world won't work if we can't find a way to bring the oil and coal people along: folks like Jim'll just end up poor white trash like the rest of us without justice in the changing of the guard. I'll raise him as well as I can in my old age once I'm out. I had a son. His name was Toby. I had another."

"His name was Brennus," Bren said. "Weirdass name if you—"

"I have a third, Bren. His name is Water Jim. Take care of him until I get out and can talk to Jim about it. It's gonna be harder than hell to forgive that man, but I have to find a way. We all do. Raise Water Jim as you would your own baby brother."

That year Marionette gave birth to two boys, back to back. Not literally back to back — a double breech birth's an awful way to go. First one boy, then another one about ten months later. Remmy didn't get to see neither of them that year in 1981, but he thought about how Beth got three grandkids in one year. Two by Mary and one by herself.

The years dragged on.

WILSON REMUS

1986

REMMY DIDN'T KNOW much about my cousins but at least he did hear about how HUGE snowstorms came through Southern Illinois and those boys got on top of Marionette and Dean's house and the snow drifts piled all the way up there and they would sled from the roof all the way down to the road. Their other grandad took the pickup, built a sled and took the boys all the way over the road, pulling them at forty miles an hour. Two wooden water skis, they built a toboggan on it and went every which way. They had to borrow Bullhorn's snowmobile that year because they couldn't get anywhere.

Well they pulled some grown men on that sled and snowmobile combo and passed this farmer's house and the dogs tore at them. Chasing and gnashing their teeth and the grown men screaming as loud as Marionette's boys had screamed. Those men ripped off the handrails from the sled and started whacking at the dogs to beat them off and away. That sled was chewed up.

The dogs followed them all the way until they got on the interstate, dragging snow.

My cousins came to him that year and so did his son Water Jim. The four-year-old snuck off into the main yard and picked Remmy a flower. It was the prettiest little dandelion you ever did see and he picked a little wildflower along with it. A guard in the tower called the warden's office. Here came the deputy with the State Police law deputy sheriff man right beside him. That old fool told Remmy, "If any child goes into the yard and picks another flower, all visits will be terminated."

"This is my son," Remmy said. "The flower pleases me."

He read *Speaker for the Dead* that year and wondered if any of them grandsons would tell the whole truth about him when he died, both the good and the bad.

I tried.

You're reading it.

Ω. Coda and Overture

...Hence, if belated drops yet fall
From heaven, on these her plastic power
Still works as once it worked on all
The glad rush of the golden shower.

— *Jack Lewis*

"Wisdom is the recovery of innocence on the far end of experience."

— David Bentley Hart

"Where does a wise man hide a leaf? In the forest. But what does he do if there is no forest? He grows a forest to hide it in."

— G.K Chesterton

"It may be that he has the eternal appetite of infancy, for we have sinned and grown old and our father is younger than we."

— G.K Chesterton

"If your boy is a poet, horse manure can only mean flowers to him; which is, of course, what horse manure has always been about."

— Ray Bradbury

WILSON REMUS

1987

First thing Remmy did the day he got out of jail was go to a baseball game right there in the city. Ten years of prison will do you, he figured. Uncle Charlie'd gotten ten years for shooting a guy in the belly and Uncle Nicholae hadn't gotten none for being a Pentecostal and maybe he should've.

He went to that ballgame alone, watching the whole city rise before him like he'd crested a hill that hid the trees behind it as a kid on the farm. White Sox VS. A's. The Athletics had a rookie named Mark McGwire who hit a home run right to Remmy. He caught that ball and thought two things:

1. *The Cubs would never win the world series. Ever. That's why you came to White Sox games. For the home runs.*

2. *He'd give this ball to his new grandson as soon as I was born.*

Turned out I was busy being born same time. The moment Remmy caught that ball, Lancelot Tobias Broganer was born to Bren and Danielle, making Daddy John a great-grandad. That's me. And he recalled my birthday: it was April 30th, the same day in 1945 that Adolf Hitler'd shot himself, Pete Taylor'd flirted with his sister, he'd defended her honor, and those men talked about the end of the war and fires you start and couldn't stop. He'd started a fire like that once. It was time to do all he could to stop it.

Remmy started heading south to find his Water Jim. Once he got there, he hugged his little boy. Sat down with Beth and Jim together. All the cockiness had bled clean out of Jim. He looked a shell of a man. Remmy'd wanted to bring a warhammer and beat the man's brains in. Beth wouldn't let him. So he'd wanted to bring a sledgehammer and she wouldn't let him. He haggled with her from blacksmith sledge to titanium ball hammer, body mechanic hammers, club hammers, roofing claw hammers, down to cross and straight pein, framing, ball peen. Nothing. She finally agreed let him take one tiny smelly old black bell hammer from that old cast iron dinner bell they used to ring. They met at a small picnic table in the park and it took near half an hour of silence, but Remmy, seeing the man all hollowed out before him, finally set the bell hammer down and passed it across the table, where it sat. Remmy waited for him to go first. Waited a long, long time.

"What… what I… what I tried to do… to Beth and, well, and I guess to you, Remmy, through Beth… that…

God you'll never forgive me for that. But it was wrong. And so was burning your doghouse. And stealing your tools. And cutting the brakelines in that car when you tried to tow it with your dad."

That surprised Remmy: explained why the towing had gone so hard, but he stayed silent.

"And for buying up the land from underneath you. Especially the fort. I saw you walking it."

Beth said, "I forgive you, Jim, terrible as it was. Man can go to jail for that. Or worse round here."

"I forgive you, Jim," Remmy said. "I… it's the hardest thing, but I think I do."

"Well—"

"Now wait," Remmy said. "You ain't the only one that screwed up. We bullied you cause your company bullied us. And it wasn't right to whip a cat on your brother's account or piss on your notebooks or none of that. Wasn't right to cover your kids in shit or make you smoke a shit cigarette."

"That was *SHIT?!*"

Remmy'd been taking a drink of soda and it shot all out his mouth and nose when he heard that. "Oh man, I'm so sorry, but that's funny. Oh soda burns the sinuses, but still it's ever funny."

"Remmy!" Beth said.

He stopped laughing. "I am though, it was mean. But it was funny. It was funny cause it was mean."

Jim Johnstone started chugging water to wash out a decade-old taste.

"But I did wrong by you, Jimmy. I did. Come with me." Remmy took him into the woodshop and taught him the way of the grain, how to work with it, and they started planning these real old red oak slabs, binding them together and sanding them. Before long, they had themselves a roundtable they'd built together, Remmy and Jim. "You know," Remmy said. "It's one thing to build a round table for my Merry Men and let Lancelot and Mordred tear it all down and make Camelot rot. But it's another to make a table big enough that Mordred and the adulterer get a seat at the table once we been reconciled. Table big enough for even the Sheriff of Nottingham to have a share in the commons: just one share, no more."

Beth had brought some slices of bread and dark cups of water, ringing bell hammers on three little golden bells, connected by one handle. Three times, ringing like that. "Take and eat," she said.

Jim and Remmy did. Turned out to be carrot cake and fine wine. "I thought you brought water and bread?"

"I did," Beth said. "Huh."

Jim lifted sad, tired eyes up to him.

"Pete Taylor died cause of me, Jim," Remmy said. "Not you. You ain't no prankster but I want you to take his place. I named my new boy after you: Water Jimmy. I want you to help me raise him. Like an uncle. Like a brother would."

Jim had nothing to say to that. He just cried softly. "You know," he said through the snot. "You keep mixing metaphors, Remmy: Merry Men's Robin Hood. Camelot's Arthur."

"Best metaphors make something new out of the old: a round table rebuilt from the splinters by Merry Men. Not for Camelot, but made new out in the wild commons of the king's land."

"How you gonna do that?" he asked.

"We're gonna need to mix a lot of land, oil, and water."

The next year, they put 30 road flares on Mary's 30th birthday cake and almost burnt down the house. By 1990, Hayden — my brother — was born. 1997 came my sister Avalon. I, Lancelot Tobias, got baptized on Easter that year the very hour Daddy John Died. That was also the year Hayden and I pulled the infamous Moon Boys prank, from which we get our nickname as brothers, but that's another story for another time.

WILSON REMUS

2005

RYAN, REMMY'S BROTHER-IN-LAW, passed away. Gwen — Guinevere Broganer — married Jim. That made them real brothers. Pete Taylor and Momma and Daddy John weren't there. Neither was Grandad Patrick. Now *Remmy* was Grandad Patrick, which was strange. He felt all out of place. It made him want to build more homes for his family. For free. Seeing Marionette happy like that. By that time, Jim'd bought enough oil wells back from Texas and Remmy had done pretty good with dividend-paying stocks. They'd worked with local farmers and bought plenty of land and together with the settlement from *TEXARCO V. BELL HAMMERS* drew up paperwork for a commons all around the old ford. Shared land that paid out quarterly dividends to all the citizens of the county: private property for one and all. They started the project with his grandkids and Jim and Water Jimmy and what remained of the Merry Men, now with his reconciled Mordred at the table. I was doing more and more theater and music at the time and did a play on Robin Hood and his Merry Men. I played a jester. Water Jim played Robin. I'd trained under the fight choreographer from one of those crazy films about magic called *Return of the King*, Remmy thought he'd read the book a long time ago but couldn't quite remember. Anyways, I learned how to broadsword fight and was supposed to choreograph this big old fight scene with swords and things so I needed about a hundred swords. So Remmy took an old fence made of those thin one-by-twos and we cut angles to make two-edged bastard swords out of all of them and then taped up hilts and painted them with the chrome bumper paint he'd used on his own bumper after the well corrosion and on Bren's bumpers. Looked pretty good and he liked the idea of something he made getting used to tell the next generation the story of Robin Hood.

But that play was nothing compared to the real thing. See Bullhorn (who was old as shit) and Norm and the Holsapples and Taylors all moved into the fort out there in King's land. So did the rest of the Merry Men. They'd built onto the castle over time, begging for stones — not money, but stones tilled up from the farmland — just like St. Francis or Bishop Castle in Rye, Colorado. They added on and made a great hall and towers and the rest, building and building with the Masons and Smiths and all them old carpenters and the laid off oil folk too (corporate consolidation and bloated CEO pay hurts them just as bad as pilfering the king's commons). In the middle of the courtyard they hung high that old cast iron dinner bell and below it they put the almost unharmed meteor and it held a red glow, almost magic-like. *Go deeper*, Remmy heard it call to him. He stared at it a long time and finally told me to puzzle it out for him. The fort wasn't paradise, don't get me wrong. Wasn't no utopia: you had to work at it, living together like that, sharing life. Someone had to take out the trash, after all. Ain't "work" one of the rules of life for monks? You had to forgive. You had to confess. You had to submit to one another out of brotherly love, but it worked the more they worked it. The first thing the Merry Men did, gathered round the table, was — with my help — to send *Arctic Projections of Worldwide Warming and their Effects on Arctic Drilling* to several places (eventually the L.A. Times picked it up). Remmy stayed out of the whistleblowing, due to the contract and court ruling and all, but we carried the fire. We'd figured out a way to hold the land and natural resources in trust for the people of Bellhammer, Illinois and large chunks of Little Egypt. Whatever sold out of the land — oil rights or land rights or even intellectual property — got put into a permanent fund and the earnings off that fund paid a dividend to every man, woman, and child who lived in the region. They gave Little Egypt its namesake back and not a moment too soon cause racists had made it even worse, maligning its good name again, this time in praise of slavery. We couldn't have that in the Land of Goshen, no sir. Little Egypt set free and restored and taken back from the robber barons and divisive men both: in the land of milk and honey, sometimes it's not the bounty but the brethren — sometimes the milk's in the driver of the milk cart and the honey's in the keeper of the bees. Sometimes the iron's in the Smith, the feather's in the Fletcher, the bread hides there in the belly of the Baker, and wine hides there in heart of the Cooper.

WILSON REMUS

2011

COPD THAT HAD started with Texarco's asbestos boards got worse with some smoke inhalation. Remmy went to the hospital. The brain fluid was off and was hurting his memory. They put a drain in. Something like a siphon to take off the pressure and help the memories come back. His heart was failing soon after. They had to do bypass on that. There were several scares. The Good Lord returned to him in that time: "Remmy."

"Oh Lord I have missed you."

"You've been running, Remmy."

"Have I?"

"I've been chasing you, Remmy."

"Why does it feel good to feel bad?"

"Because you're trying to build what only I can build, Remmy. You cannot build a place of love by yourself."

"I couldn't save them. None of my buildings. None

of the things I made. I couldn't build it big enough, fast enough, safe enough, welcoming enough."

"That's my job, Remmy."

"That's your job, Lord." He was crying again. He felt like he'd spent half his life crying and the other half laughing and didn't have no between times. Beth passed away that year. Heart attack got her in fifteen minutes. He looked for Bren but Bren could not be found. Bren'd threatened to go to Brazil — maybe that's where he'd gone. Police found Bren's front door wide open and the lights on and all the stuff in there sitting and coons and groundhogs in the kitchen eating all of them boxes of fake junk sugar flakes them big old companies sell as "breakfast." But Bren could not be found.

Kids all stayed with Danielle, of course, what was left of them.

Marionette came to him then with her husband Dean. And they sat and he looked in her eyes, her old hurt eyes, and he said, "I've been working on a building project."

She scoffed. "Yeah, what's that, dad?"

"I'm going to rebuild my relationship with my daughter. You ain't no puppet. You're my little queen of heaven."

WILSON REMUS

2012

By the time 2012 rolled around, I came home as often as I could from Joplin, Missouri to start interviewing him as to the stories of his life. I tried to stick to his way of telling things, with no care for how I think or what I've read. I think I told it true, even if it was just bullshitting, but it was hard getting him to talk.

"I don't have any stories prepared," Grandad Remmy said to me.

"Well we'll just talk about whatever comes up," I said.

"Well sometimes it takes a while for stuff to come up," he said. His memory was failing, you see.

"I'll go fishing then," I said.

"Well you're gonna have to drag down in the bottom," he said.

"Let's go to the bottom, then."

"All the way down deep," he said and then as if he

was recalling that very phrase he said again, “all the way down deep…”

We started to record them one and all.

That year, old Remmy turned seventy-seven. The Jews would have said something special about that — a sabbath of sabbaths, twice lucky, whatever. Remmy spent the morning insisting that he was scheventy scheven cause that’s how his German other-grandpa used to say it and they all about had enough of him saying scheventy scheven. He wore his PJs. His buttonless-loose-fly PJs around the whole house until noon and refused to get dressed cause he was scheventy-scheven and he wouldn’t be buried for his eternal rest in his PJs so he needed to get the mileage out of them while he could.

Until the doorbell rang with the wrong wringer: some Irish drinking song reserved for St. Paddy’s. He said, “Dammit.”

His granddaughter-in-law had come to the door.

He opened the door and the all-glass storm door. And his respective horse was hanging out of its respective barn.

She looked down. Blushed scheventy scheven shades of pink.

Remmy said, “Whoopsie.” And tucked him back in there. “Older he gets, the less he minds. Kinda like me and PJs.”

She asked, “Is—?”

“It’s my birthday!” Remmy said.

She said, “I know. That’s why I wanted to sing. *Happy Birthday to you! Happy—*”

“No, no, come here, come here.”

She followed, timid.

Remmy said, "I don't need you to sing to me, I got my singer ringer." He hit the test button on the big cedar-encased doorbell. It rang the St. Paddy tune. "Dammit." He pressed it again. Auld Lang Syne. "Hold on."

She smirked.

The twelve days of Christmas started. It was very, very long.

"MARY! How the hell you get it on Happy Birthday?"

The ringer rang *FIVE GOLDEN RINGS.*

"Oh Dad, would you quit it with the doorbell? We're getting things—"

The ringer rang *SEVEN SWANS A SWIMMING.*

"But this is important!"

The ringer rang *FOUR CALLING BIRDS, THREE FRENCH HENS, TWO TURTLE DOVES.*

Remmy said, "AND SOME PARTS TO A MUSTANG G.T.!"

It continued ringing while his granddaughter-in-law cocked her head at him. "Mustang G.T.?"

The ringer rang *NINE LADIES DANCING.*

"That's the Little Egypt lyrics. You didn't know that one? MARY!"

"Good God," she said.

"Close enough," The Good Lord said.

"She's learning," Remmy said. "I'll learn her."

"Teach her to listen first," The Good Lord said.

"It's hard with eleven pipers piping in your ear," Remmy said.

"What's hard?" Water Jim asked.

"I'll say," Mary said.

"I didn't start the doorbell," The Good Lord said to Remmy alone.

"Yeah but it's your song," Remmy said.

"The season's my season, but I have nothing to do with that parade of Victorian boons."

"That where we get the word boondocks?

"What?" His daughter and granddaughter-in-law and now his grandsons like me and my wife and great-grandsons like my second cousins all said, standing around the great harold of seasons out of time and holidays that would never be.

They cut it off on the twelfth day.

Remmy got through LOVE ME TENDER, JINGLE BELLS — which he pointed out was a *Thanksgiving* song — WINTER WONDERLAND, THE RITE OF SPRING, SUMMER LOVING, and DANSE MACABRE before he got to HAPPY BIRTHDAY and the family all but the granddaughter-in-law retreated.

"See there, don't need you to sing to me. I got a doorbell."

"Well yeah, but at least I know the gap between *The Rite of Spring* and *Danse Macabre*."

"I don't see much gap between dead seeds and saplings," he said, "one flows right into the other."

She didn't have much to say to that.

"Where's the D.S. so this piece of music will stop?"

They all forced him to change into his slacks for his birthday, which he normally would have done as early as a British WWI general and nobody thought of the hypocrisy of having never submitted to his dress code themselves, but

I suppose cared only for being spared the presence of his respective horse.

After lunch, his granddaughter offered him a slice of chocolate ice cream cake. He ate it. He said, "That piece is too small!"

She brought back a bigger one. The biggest one she'd cut.

He ate it. "Still too small!"

She brought back the whole cake and put it on his lap with a metal spatula.

"That's my girl," he said and dug in.

"Dad!" Mary said.

"Grandpa!" someone shouted. "You'll get sick."

"Dad you'll get your pants dirty!"

"Pants wash. Pants rot. If I'm worried about one pair of pants at scheventy scheven, I got serious problems." He ate on and finished that whole ice cream cake.

"All that ice cream. Ain't your teeth cold?" Mary asked.

"Dentures got their perks."

Soon after that, me and this draft — or one like it — and the sample from the meteorite's core went to New York City. I came back for month-long stretches to work on the houses.

WILSON REMUS BROGANER, 1935 - 2015 ARCHITECT OF TIME

HE AND HAYDEN and Water Jim and I finished the houses together. Then we'd drive him into town to play poker and he'd pay for our gas and meals and beer and said, "When I was your age, I played buckkeeper with Grandad Patrick and he never gave *me* beer." He laughed a wheezy laugh and got into one of his coughing fits.

Marionette and Dean used a good chunk of the money they'd made on their pharmacy monopoly to build a mansion and buy a lake and a hundred acre wood just like in Winnie the Pooh. I sold shares and Remmy sold shares. Remmy did one more, though, see. One day he while he was scrounging up money to build his paradise farm, the Good Lord said, "You remembered me, Remmy."

"I did, Lord," he said, "though I didn't obey you very well sometimes."

"Already forgave you for that. Now remember."

Something like a hammer struck something like a bell in his mind for a third time.

Remmy remembered the place. It was a place he'd marked with his toe or a flag made of a stick and a rag or had written his name in the dirt there with his piss just to make sure he still knew. There'd been a white brick restaurant and the railroad. Truck stop there. Turned right into a pants factory for the war. Well before all that, the big warehouse that'd done the bottled cola there: Spur Cola from Bellhammer, Illinois? Remmy'd watched that plant close one day in the war and watched them take all of those bottles — just a bunch of them — and he'd followed them out and saw people dump them into a specific mine shaft. And "One Day" had finally come. He hired a big old earth mover from Fabic in town and he dig up all those bottles cause there wasn't another Spur Cola in the world but in Bellhammer, Illinois and therefore those bottles would be prime rare antiques. He told Aunt Mary, "Hayden showed me this thing called eBay and Lancelot showed me another one called Etsy. First one's like having your own auction house without the overhead. Second one same thing, just an antique store." He and Water Jim sold every last one of those bottles and with all that and his shares (as well as some mystery sales bins he'd bundled for folks on them internet stores), he started building for all of his friends and family that was left. He built a little house for two old bachelors and a nice loft for my brother Hayden to visit us. I learned a lot building that house with him, Hayden too. Grandad Remmy took his notes on studs in shorthand. I said, "What's that?"

He said, "Shorthand. I'm taking notes about the north-side of the house — things I forgot."

"What's shorthand?" I asked.

"They don't teach you shorthand in school?" he asked.

"No," I said.

"Dammit I knew nobody was gonna write like this in the future. And here we are in the future and there ain't nobody writes this way no more."

I took my notes and the thing I'd pulled going deeper with the meteorite like he shoulda done and I took off to New York to finish the job that needed be done.

They stocked the lake with fish and aerators for the fish and planted heirloom seeds and all the woods was overrun with whitetail so bad in Southern Illinois that the state turned wildcats loose though they denied it, wildcats like the mascot of Salem's high school. They had chickens and they had their friends buy up nearby farms and they stocked up on canned goods and ammo because there were a few preppers among them, but you know, Remmy didn't mind because they were there and they were together and he had rest from his labors.

Here's what else I know as sure as shooting skeet: there's this thing called resonant frequency. I hear everything's got it. It's the thing that'll get a piano string to vibrating when you sing the right note. It's the thing that'll get a pot to make its sound when the soapy water swishes in it just right. Everything has a pitch, you see. In the right mood and place and shape, I bet you could even get a dry old turd to ring out some sad and beautiful song that'd break your heart if you struck it just right and found that pitch deep inside of it that got it to tell you how lonely and unloved its life had been, how it'd been the butt of every joke. How even the worst, most rotten, alfalfa-smelling piece of shit has a story the Good Lord alone can tell. How the literary snobs of them big cities like New York look for some great symbol when it's sitting right there in their own toilet. The folk of Southern Illinois have told off religious hypocrites and snobs and rich folk like that before saying, "Don't you act like your shit don't stink." But still rich elites and church ladies call places like Little Egypt the breeding ground of white trash and deplorables and cucks. White trash children just like Water Jim, born of a tryst in a prison cell between an oil person and a prankster carpenter. They call them shit-towns. They call them shitholes. They call the people that live there redneck pieces of shit and make my family's lonely and unloved lives the butt of every joke while poking fun at the country songs that make you cry, calling them naïve and kitchy. But my little pieces of shit sing beautiful sad songs of lonely laughter. My pieces of shit gave birth to me and taught me to sing. And without my pieces of shit, the guy or gal who bought this book never would have learned, never would have made a dime off of its sales. Even if that one singing turd was all there was is in the middle of just nothing, just nothing *as such*, just plopped and festering there in the middle of the abyss, that turd would tell you everything there is to know about all there is and ever was. You could get a turd to sing the song of creation: all that is earth has once been sky. It's true: you strike anything right and it rings out the songs of angels. The Broganer clan, far as I can see, were those strikers. They'd hit everything just right and get it to smile or weep. They spoke the language of ambience and sang the music of the spheres. They'd get a murmuration of starlings to stop dancing and listen to their tales. They'd get a woman to give birth right on time with their work and wonder. They'd get a stove to work right for listening and a land to give water with nothing but the right joke and a short handled shovel. They'd summon groundhogs from the pit of hell and call down scaffolding from the pearly gates. They'd match the resonant frequency of every stock and stone and loaf of bread and glass of wine. They were, and are, bell hammers. They strike at the deep places of the world and each thing that stands up under that hammerfell... well didn't that one guy say once that each thing finds tongue to fling out broad its name? Takes a bell hammer to do it right and a Broganer's as good a hammer as any, strike they pearly towers, common turds, the bastard child of an oil baron's rape, or swinging they blindly in the dark to break, break, break at all our terrible secrets in hopes to shatter gold.

When Remmy died, he died of cancer caused by all that poisoned salt water.

Yes, when he passed, he was walking out around the lake on the northside, passing the crops about to be harvested, hiking up over those fallen logs and the four wheeler path, and he laid down there on the clover of his daughter's estate, tired and giggling.

He went through dark and death and the abyss.

On the other side of all of that, he met the grounds of all reality.

"Well hello there, Remmy," The Good Lord said.

"Hello there, Good Lord."

"Now Remmy, you know that I love the joy you have and the spirit of play I put in you."

"Yes Lord, I'm an ornery son of a gun."

"Well you're my son first, Remmy, not a gun's and there's a difference between having fun with someone and having fun at someone's expense."

"Ah."

"Now I invented laughter. I told some good jokes when I walked the earth."

"I love the one about the plank and the speck."

"You always did like slapstick." The Good Lord smiled. "But did you have to give them all lice?"

"Oh come on," Remmy said. "Finding them adult lice in my head was the best chance an old poltergeist like me could ever ask for! I knew I was dying so I just went and tried on all of my hats and scarves and Mary and Dean's hats and scarves and—"

"I saw it, I saw it all," The Good Lord says.

"Why it's a funeral full of lice!" Remmy said and he laughed. "Didn't you make the lice too like them spiders and groundhogs?"

The Good Lord sighed and shook his head. "I didn't make them to hurt people. And I didn't make them so you could let them hurt people just for a laugh."

"Well… I suppose that's true, but it's still funny."

The Good Lord said, "Danielle and the rest of the health department are gonna have work for weeks cause of that one."

"Job security," Remmy said.

"Oh they were plenty busy. They had vision and hearing screenings that'll get bumped to the end of the week."

"But them school kids can go a day without glasses," Remmy said.

"Yes, but then that'll bump the immunization shots to the following week."

"Well shots hurt! I'm helping."

"Everyone except Jimmy Hunter. He'll go without his tetanus shot, will be barefoot out in the yard, will step on a nail, get tetanus, have the muscle spasms, and die."

"All cause I made the lice worse?"

"Well there's other things too, it's not just you, Remmy, but that's what I mean by fun at other's expense."

"I see."

"Do you? *As the man who shoots fiery darts into the air not knowing where they will land, so is the man who says, 'I was only joking.'*"

"Now that ain't fair, you going and quoting scripture at me, Good Lord. You know way more of the Bible than me."

"Remmy? All I say's scripture."

"I see."

"Do you? Do you get that all tyrants start out making fun of people and nitpicking and showing how they're better than someone else? How quickly a prank can turn to bullying? How quickly bullying can turn to a lynching? And how quickly one lynched people wants to go and lynch their lynchers? How's that kind of myth of redemptive violence plays out over the rise and fall of nations, Remmy? A tyrant's just a guy who's a pro at having fun at other people's expense and it's poverty that a child dies so that you can have the fun you please, Remmy, don't you get it?"

"I do. I think that's how Jim Johnstone got to be so mean."

"Peeing on his notebook didn't help him much."

"I see. Good Lord I was an asshole." He laughed and shook his head. And he cried. "An asshole to everybody."

"The space between you and Jim Johnstone ain't that far, Remmy. It's only a couple of decisions thick. The only difference is that you listen to me now and again when you do something mean and let me change your mind. Your best jokes were always inside jokes with you and me and didn't matter whether anybody else laughed or not. Wish you'd have listened more."

"I see."

"You do," The Good Lord said. "I'm glad."

"Will you let me walk with you and try again?"

"You'll do, Remmy. You'll do. I can build a lot on a soft heart like that. Have before. Will again."

"Say, I'm dead, right?"

"You've returned to me until the full number come in and I give you all new bodies."

"Like a resuscitation in an ambulance?"

"No. I'm the creator."

"You're gonna make a whole new world?"

"That's the idea. Made it once, didn't I? Can't I make it again? Can't I write an ending to the song that takes all them bad things people ever did and all them bad verses people ever wrote and tune it all true?"

"I'll be damned."

"Not on my watch. Remmy, am I good?"

"You are, Good Lord."

"And am I great?"

"You are, Good Lord."

"Then don't you think I'll make the right call on

everyone and make it alright in the end? Wasn't the best parts of life that I made good in the first place?"

Remmy thought. Then nodded.

"You want me to go get Beth?" The Good Lord asked.

"Not just yet. Can I just have some time with you?"

"Time's not really a thing here, Remmy. That'll make more sense as you look around here. Toby's here too if you'd like."

"Oh I'd love that," he said.

Toby came. Toby was grown. Toby looked identical to his grandson. They talked for a while and then Toby had work to finish so he went off.

Remmy said to The Good Lord, "You know, all I ever wanted to do was to make paradise."

"Paradise is more than a place. It's an architecture of time, Remmy. And only the maker of time can build that: you're not strong or wise enough to build eternity."

"Ain't that God's honest truth."

"Literally," The Good Lord said.

"Thanks for making me, Good Lord."

"It's my pleasure, Remmy."

"Really?"

"Oh you give me great joy, yes my boy. Even when you fail."

"And Good Lord?"

"Mmm?"

"Thanks for making the song of bell hammers too."

The Good Lord grinned wide.

"Thanks for making it so that on a hot day, the snap and knocking of the hammerfell on a brand new roof will echo into the trees of town and off the houses and down the lane. Thanks how it works out like nature's drum line, like mankind answering the call of the red-headed woodpecker with four hammers and a song, the drumming of nails, the high siren of saw blades, the groaning of the old oaks growing around their fallen timber and the way a good lumberjack will harvest only those old ones. For the harvest of dying trees, I thank you. For the seeds of saplings, I thank you. For propping up Sister Earth this long, I thank you kindly. For keeping Southern Illinois alive and drumming even after the last tree in Shawnee falls."

"Ain't trigonometry beautiful?" The Good Lord said.

"You can do all that with time and space?" Remmy said.

The Good Lord smiled. "Why don't we go sip some sweet tea and sit at the end of this pier over here. Kick our feet out over the waters, barefoot like. For your sweet tea I got twice the ice, but it's free now, all free. I'll show you how to walk on them waters if the urge ever takes you."

"I'd like that."

They went to the pier. It was built out of old milk wagon boards, the paint chipped from the old names of the old dairy farms that delivered milk. There were Lincoln Logs. The Good Lord sat down and passed the sweet tea and said, "Take and drink." Then The Good Lord starting building castles out of the Lincoln Logs with Remmy, sipping sweet tea and listening to the loons cry Remmy back to sanity — listening to the loons cry Remmy back to what the ancients called a good humor. "You've been wanting to come home for a long time."

"Yes sir. Thank you sir. I've needed it something fierce, sir."

Three bells, just a ringing on out there as one until their

sounds perfectly synced up together, cancelled each other out, and turned to thick silence, a weighed silence like a measure of love, a countermeasure of the loved, and the measure of love between them, a silence of all that light and meaning hiding out there in the folds of outer space, like a body there in the bread, like lifeblood there in the wine, the song that rang out every time the up crossed any across, every across looking up sharp like a cross. Home in the outside. Adventure on the inside. Remmy looked up at the stars: not over and out, not even down and out. Them stars was up and in. Towards the He who predicates all. Towards He who is the end of every start and the start of every end.

"Well come on then."

"Yes, Daddy."

ARCHIVER'S NOTE

You may be wondering, of course, how I know that conversation happened since it happened after his death. Since I wouldn't have been able to interview him then.

That's a great question.

And it's a story for another time.

END OF TRANSCRIPT

WHAT: Wilson Remus Broganer's grasping at the threshold to the abode of light, his fight against a Clark Brothers subsidiary, the redemptive clockwork of Elizabeth Brograner's struggle as an oil person and of Jim Johnstone's reconciliation, and the first Broganer attempt at ambience.
WHEN: 1935 - 2015
WHERE: Southern Illinois, America [/go], Earth
HOW: Third-person narrative through interviews, accompanying data, research in the SIBL branch of the New York Public Library and archived consciousness.
WHOM: Archivist Tobias for those stacks preserved by the Aurelian line.

THE MAKING OF BELL HAMMER AND ACKNOWLEDGEMENTS

Some books happen on purpose. Some books happen on accident. This one's an accident and it's the best accident that ever happened to me. About seven years ago or so I started recording advice from the men in my family partly because I wanted a sort of record of things we needed to remember — there's not much institutional or family memory in Southern Illinois in some ways — but partly because I really had never sat down to formally pick the brains of my grandpas and my uncles. I turned on a recorder and started learning things from everyone that no one else knew about. Tara, my bride, also started sneaking recordings in the middle of conversations.

I found out, for instance, that my Great Uncle Kenny had been in *both* the 101st airborne and the 82nd airborne in a behind-the-enemy-lines tank buster squad that first ran out of grenades and then found an old welder. So they started welding the lids of tanks shut until they snuck back toward the allied line of defense. There's probably a novel in there in and of itself, certainly there's enough newspaper clippings, but crazy things like that came out. I learned about pranks my great-great grandad had pulled like the cross dressing T.A. Or the carrot in the fly gag. The list went on and on.

I interviewed my quiet grandpa second and that, in some ways, was a mistake: he passed away two months into interviewing him about three years ago. I got my ass in gear and finished recording as much as I could from Grandpa Jerry, adding in stories from my bride's grandpas and the rest and suddenly felt shocked to realize just how many times my family kept saying, "Yeah, we shoulda got an attorney." For great things and little things, but all of it was too expensive for them. And most of it centered on Texarco.

Then I read the climate report that came out of Exxon: that Exxon not only knew about climate change but *planned* on it so that they could drill deeper under ice. And the borders of the story started to form.

They say write what you know and as someone who's inherently motivated by science fiction and fantasy, I find that a silly idea. But the moment I started, all of these deep roots and stories and moments came back while living here in Southern Illinois. I'm not much of an iceberg writer like Hemmingway, but I can tell you confidently that for the 90,000 words in the finished book, there's another million words left untranscribed, unrecorded. So many stories. You want another Remmy story?

Well one time Remmy had to have a bunch of Freemasons and Shriners over for dinner that he didn't like all that much but he had to on account of how much influence they had in the town, he just had to do it. So he bought up some old chairs and took out the planks underneath and reupholstered them with some old curtains he'd found out by the side of the road. Well they all stood respectfully when Beth came in, all around the table and Remmy said, "Let's stand and say grace standing." And he said, "Thank you, Good Lord, for humbling us like you do and dragging us low so that we can know what it is to suffer well like your son." And then he said, "Let's be seated," and all ten of those Freemasons and Shriners sat down on those chairs that had nothing but fabric and they all bust right through, butts and all, dangling out the holes like some sort of barreled boys.

True or false?

Did it happen or not? Funny thing is that people almost always get it wrong. Almost always. They think the fictions happened and they think the true stories are made up, so what does anyone know about the historical "our world" sequence of events?

Does it matter?

This is the problem with stories and everyone I could acknowledge, at some point or another, would call me an exaggerator just like my grandpa. "Why are you lying so much, Lance? You insecure?" Well of course not, that's not what fiction is. That's not what fictions do. Journalists just don't get this and there's no room for good bullshit anymore. I don't mean in the newspaper. I don't mean in data for your stock trading. I don't mean accurate measurements for a house or a healing femur. I mean stories that convey the truth of how it felt to live through a time or do a thing. The fisherman's absolutely in the right to exaggerate the size of his fish: almost all of the time, he's not concerned with impressing you with the size of the thing. He's more concerned that you feel what he felt when he caught it. And that's true. That's empathy. That's what this is all about.

And our grandpas got it. And they were better at it than me.

You start out wanting to write four generations of carpenters interwoven with their years in parallel. Had I stuck on that track, the book would have been 400,000 words long and thank the Good Lord that authors Lisa J. Cohen, Amy Rachelle, Therese Walsh, and Heather Webb saved me from that madness and encouraged me to focus on Remmy. Donald Maass's Socratic questions at the Writer UnBoxed Uncon and subsequent beta read saved my bacon as well. I drew from a deep freshwater well of encouragement from the folks I dedicated the book to but also unsung heroes like the entire Lang family — Ellie, Aanna (and Logan), Emilie (and Eddie G… who brought along Cousin Vinnie), Abigail (and longtime fellow Night's Watch brother, Robb John Kimball Jones), Esther (and Jeremy), Eden, Emma, Gavin, Jessie, and their cousins Joseph and Susanna — who in a more or less unified voice have spoken into my craft through their unique heart languages. They all continue to encourage me to write probably more than any other given family and that includes their early and patient collaborations with me in my ever-fumbling attempts as well as their even longer suffering through my over harsh criticism of their own work (work that far outshines my own). I felt welcomed and encouraged by all of you, something like a "Laurie" to your "Little Women" (sorry, Joseph, Gavin, and Jessie — it's the best metaphor I have and I know, Joseph, it doesn't accommodate scripts well), though not in any romantic light. (The romantic side of Little Women I reserve for the Balu clan and specifically my bride, Tara who's also writing and probably the best reader I know outside of Dr. Cirilla). Anything I write is predicated on the various forms of writing groups I had with folks like the Langs, Seth Caddell, *Lamb and Flag*, and the Limners. There are also a handful of professors and teachers that (save but the aforementioned teachers in the dedication) became the first to properly encourage me around this time. They include Mr. Waller, Mrs. Marsh, Dr. Chris DeWelt, Professor Buckland, Dr. Bowland, Dr. Doug Welch, Dr. Ragsdale, Stan Wohlenhaus, Dr. Mark Scott, Dr. Woody Wilkinson, Dr. Mark Moore, and the aforementioned linguists.

A story grows and it gets passed to beta readers like Mark Neuenschwander and Peter Corado and Colby Williams and F.C. Shultz and Matt Otey. The constant advice via text from Rev. Kyle Welch, Dr. Thomas Alexander Giltner, and Dr. Anthony G. Cirilla. Then Zach Spiering, Jessie Weiss and husband Zach Brazle (who sacrificed a front incisor to figure out that you can't hold a Beretta pistol with your teeth), Justin Friel, Isaiah Bayse, Jenni Reilly, Nina Chung, Chelsea Grasso, Jessica Drake, Adam Jones the world sailor, Logan Stewart, Gregg Hull, Joey Hawthorne, Jenifer Slabaugh, Karl Mitchell, Jessica Scheuermann, my bride Tara Schaubert for putting up with my long belabored season of familial and geographical myopia, the Redeemer writers and readers, the Sirkman twins and Rabbi Sirkman, the various NYC reading groups, Emily Munro, The ACT International staff, The World Fantasy Bears, Kaaron Warren, over a hundred longtime supporters and patrons, the early ARC readers at NetGalley and Voracious and Booksirens and Goodreads, and readers like you: the list goes on and on. Eventually it ends up on the desk of Dave King, who edits it as close to perfection as any human is capable of making my struggling attempts. Paul Brown, the PBS documentarian, for showing me how much the issues parallel those in Alaska — and for the Hickel family, Jack specifically, enlightening me about the GLOBAL COMMONS. Gosh, I can't even remember who all helped with this one: probably a hundred people. Probably a zillion. I'm grateful for all of them..

I'm grateful to the families of oil workers who helped round out my picture and gave room for Water Jim, many of whom I met with Paul Brown — friend late found, mentor for life — while on the documentary survey in Alaska.

Really, if you're reading this, I'm grateful for you. You gave a chance for my family to have a voice when, up until now, they really didn't have one. And that's all a man can ask for: the seventh generation is the storyteller's. This is our tale.

This is yours.

It's up to you that others read it. Please, once you read the book, sign your copy and pass it on to a friend. I love to see these things out in the world, changing hands.

A LITTLE EGYPT GRAMMAR

I fought harder with the grammar and syntax of this book than anything I *ever did write, it's true as tinsel*, as Remmy might say, but the final version is far, far tamer than my first attempts at catching the Southern Illinois dialect as kingfishers catch fire.

For starters, I have Dr. Larry Pechawer, Dr. Kenny Boles, and Professors Dave Fish and Chad Ragsdale (not to mention my good friends Dr. Thomas Giltner, Dr. Anthony Cirilla, Dr. Stephen Lawson, and Dr. Jordan Wood) to thank for helping me even consider the grammar, syntax, and dialect of Little Egypt. They are the real grammarians in my life and where I attempt to put into practice what they know, they are the ones who have mastered Hebrew and Latin and Greek and Syriac and Old English and the rest. I their foil, they my diamonds.

The truth is, after talking with them for a while and recording my grandfather, father, and the fathers and grandfathers of my friends, I began to believe that — functionally speaking — verb endings are on the rise again in Little

Egypt. It's true that they might never get written down and formalized as they once were in Old English (were they *really?*), but I would feel it a crime to withhold my homework. My sophomore Algebra teacher scolded me often for doing as much.

In the first version of this book — indeed in one of the excerpts that sold to a literary magazine — I found myself using apostrophes as often as Mark Twain but never in the front of the word. Suffixes worked to capture their tongue, prefixes did not. And heaven forbid I invent my own version of a word like *fo'c's'le deck*. Ultimately, I concluded that apostrophes get in the way far more often than they help and I have both Cormac McCarthy and Reverend Kyle Welch to thank — McCarthy for his hatred of punctuation and Welch for his explanation of the phenomenon. I feel now that turning *because* into 'cause really rips the reader out of the narrative, for one, and it's also a way for the author to say, "I too went to college and I too know that the word *be* has been removed." I now feel about dialectic apostrophes the way that Vonnegut felt about semi-colons (*transvestite hermaphrodites representing nothing other than that you went to college*). Besides, not only do they rip the reader out and show the author patting his own ego on the back, they actually miscommunicate. *Because* does something different than *cause. Because* means "for the reason that" or "due to the fact that" as in "by cause," meaning, philosophically, efficient cause. I ate an apple *because* I was hungry means that hunger was the efficient cause of my eating. But when Southern Illinois says "cause," sometimes — philosophically speaking — they mean final cause. For example:

"Remmy! Why in God's name are you walking barefoot on the dirt?"

"Cause feet were made for touching soil."

Meaning that the final cause of feet is to have contact with the ground, which is true: that's precisely what a step is. It's not a proof or reason or justification so much as a statement of the ends of the being of feet. They're saying, in a roundabout way, "the end of feet is to touch the ground." There's also formal cause that they mean:

"Remmy! Why the hell are you pissing in a bucket?"

"Cause it'll hold about four gallons and that's what's in my bladder."

Meaning that one formal cause of a four gallon bucket to hold about four gallons, which is true. It's not proof or reason, again, so much as saying, "The stuff in my bladder is a perfect fit for the capacity of this bucket."

When you consider this, you'll realize that the "common sense" to which England and Scotland once alluded is actually alive and well in the daily wit of the "uneducated" class of rural America. In fact, I would argue that their philosophical pedigrees — when wit is deconstructed — can handily unman what passes for philosophy in this day and age. "Cause" is just one example of this, but I often eliminated commas and periods for the same reason. A rambling sentence will seem like it does not predicate and will actually often hold the kind of philosophical weight of the old German theological tractates. A great example is in the groundhog gauntlet:

"But that's both the bad and the good," Remmy said. "Not just the bad that can happen will happen the good that can happen will too and if you happen upon bad things like tonight and in that moment you choose to happen your good on them bad things more good than bad'll happen in the end."

That's precisely the kind of long-winded thing my grandpa said all the time to confuse people and then he'd laugh and slap them on the back and say, "Just like sex" or whatever and they'd laugh and carry on. But he meant it, and that's the key: *he* knew what he was saying even if you didn't. And then he'd shout I DON'T KNOW just to keep you guessing.

As for endings, Little Egypt will take the tense, voice, and mood of a word and jam them all into the ending and sometimes include prefixes. Some of this I kept, for instance I made no distinction between *it's* meaning "it is" and *it's* meaning "it was," letting context alone determine the tense. Greek and Hebrew grammarians will recognize this: it's alive and well in my home county.

However, the tongue of my hometown doesn't stop there. People will say something like "were'n" meaning *were in* or "took'f" meaning *took of* or even "back'm" meaning "back from." I doubt I could pose to give a full breakdown, but here's an example of a declension chart:

Tree	
Nominative	tree
Genitive	tree'f / tree'm / tree's
Dative	tree'n / tree'ū / tree'fr / tree'th / tree't
Accusative	tree'h

That is to say instead of saying "the tree of Charlie," they'll say, "the tree'f Charlie." Or instead of "the tree from the nursery" they'll say, "they tree'm the nursery," often even smashing up the definite article with the locative or dative noun. "Tree'n" is used across the board for in, on, and several different conjunctions. The "h" is more of soft schwa sound, the upside down "e," meaning that the soft "uh" is not always a place holder for thought, but rather the indication of the end of a sentence and therefore far more communicative in informal relational speak whereas formal speak sees it as a sign of ignorance. It's not. It's punctuation to the same degree that "he said" and "she said" in McCarthy is quite like classical versions of quotation marks — that and the paragraphs. For example, with tree'h rendered "tree uh"

"Jerry took'is little truckie and drover right'n the tree'h. Well Jennie, she went and came out'f the greenhouse and said, 'Not again'h.' Jerry said, 'Every damn Sunday: it's a tree'fr Sunday or a tree'th Sunday or a tree't Sunday, any which way. Maybe that's why Jesus hung on wood."

Or something to that effect. Think of it as words as punctuation — it's an oral society, mainly, so sometimes words do what silence can't.

For verbs:

To Love

Present	love
Passive	mloved
Imperfect	s'lovin
Imperfect passive	ustbeloved

Subjunctive	oudalove
Subjunctive passive	oudabinloved
Perfect subjunctive	mabinloved
Future	llove
Imperative	belove
Future Imperative	elbeloved
Aorist	loved
Perfect	vloved
Perfect passive	vbinloved
Pluperfect	dloved
Pluperfect passive	dbinloved
Future perfect	llavloved
Pluperfect subj.	oudavloved
Pluperfect pass. Sub.	oudavbinloved
Infinitive	tlove
Pass.	Tbeloved
Perfect act.	Tvloved
Perfect pass.	Tvbinloved
Future Pass. Infi.	Tvbingonnabeloved

I don't want tbelabor the point, but you can see at a glance why this didn't communicate with most urban, educated audiences or even with Southern dialects a few states over: it's actually a different language in the same way that Romania and Rome speak different languages that descended from the tongue the ancient Romans spake. Part of the urban / rural divide is that rural folk speak a relational tongue — of one thing related to another — and urban folk speak a formal tongue — of how one thing affects another in the polis. Part of it too is simply what happens when you blend Gaelic syntax with Yiddish in the riverland. Translation alone could solve a lot of things.

Obviously I'm no linguist, so I can't say with exact precision how these nouns and verbs decline, but I believe the tongue of the rural English speaker in the Middle West deserves a second look for the same reason that we do not assume all of speakers in a comparable land mass — that of Europe — all speak London English. In fact, there are communities in northern Arkansas that speak the grammar of Shakespearian English almost verbatim, just with a hick accent, as many of you know. But for now, the easiest way to get a taste of this is to get ahold of the audiobook that I read myself.

This story is part of a series of independent stories, songs, and cultural artifacts taking place in Lancelot Schaubert's universe, The Vale.

The short stories so far each have an audiobook version, in order:

1. *The Blissful Dreams of Long Ago* (contemporary)
2. *Wombrovers* (fantasy)
3. *The Blimps of Venus* (sci-fi)
4. *When Timbers Start* (historical)
5. *Earth Swallowed* (magical realism)
6. *Wilderness* (mystery)

7. *Carry Cannons by Our Side* (fantasy)
8. *Portrait of the Nonartist* (contemporary)
9. ∞ Falls 1 Stand (sci fi)
10. 160 Lug Nuts (sci fi + court drama)
11. 16oz (magical realism)
12. 2π28.57mm (horror)
13. Cast (literary)

Also by Lancelot Schaubert:

- *All Who Wander* (album)
- *Cold Brewed* (photonovel with Mark Neuenschwander)
- *The Joplin Undercurrent* (photonovel with Mark Neuenschwander)
- *Open* (short film with WRKR and MD Neely)
- *Life After Aesthetics* (poetic apologetic nonfiction)
- *Of Gods and Globes I & II* (fiction anthology)
- *Inconveniences, Rightly Considered* (poems from my 20's)

See full list of nonfiction, poetry, fiction, and transmedia published works online. You can also contribute your own fiction, poetry, essay, art, and academic work to The *Showbear Family Circus* by going to our *Submittable page.* Or just get some *free resources* for reading, writing, and making better culture by clicking over to *lanceschaubert.org*

CPSIA information can be obtained
at www.ICGtesting.com
Printed in the USA
LVHW091920161121
703491LV00016B/479/J